The
Regis Touch

The
Regis Touch

Million-Dollar Advice from America's Top Marketing Consultant

Regis McKenna

Addison-Wesley Publishing Company, Inc.
Reading, Massachusetts • Menlo Park, California
Don Mills, Ontario • Wokingham, England • Amsterdam • Sydney
Singapore • Tokyo • Mexico City • Bogotá • Santiago • San Juan

Library of Congress Cataloging in Publication Data

McKenna, Regis.
  The Regis Touch.
  Includes index.
  1. Marketing.   I. Title.
HF5415.M38   1985   658.8   84-28374
ISBN 0-201-13981-2

Cover design by Marshall Henrichs
Cover photography by Sharon Hall
Drawings by Ned Williams
Text design by Betsy Perasso
Set in 11 point Life by Walker Graphics

Production Coordinator: BMR, San Francisco, CA
Copyeditor: Ira Kleinberg

ISBN 0-201-13981-2

ABCDEFGHIJ-DO-898765
First printing, March 1985

Contents

Figures   vii

Preface   ix

Acknowledgments   xi

Chapter 1   The New Marketing   1

Chapter 2   Dynamic Positioning: The Cornerstones of the New Marketing   11

Chapter 3   Product Positioning: The Four Golden Rules   35

Chapter 4   Market Positioning: Five Ways to Gain Recognition   51

Chapter 5   Corporate Positioning: There's Only One Thing that Counts   89

Chapter 6   Developing a Strategy: Three Steps to Success   97

Chapter 7   Why Marketing Plans Fail: The Ten Competitors   127

Chapter 8   Putting Ideas to Work: Marketing Macintosh   149

Index   175

Figures

Figure 1   Dynamic Positioning   17

Figure 2   Key Dynamic Factors   22

Figure 3   The Environment Defines the Product   30
               Business

Figure 4   1977 Personal Computer   32
               Environment

Figure 5   The Infrastructure Development of   63
               the Personal Computer Industry

Figure 6   The Infrastructure Development of   64
               the Microprocessor Industry

Figure 7   Strategic Relationships   72

Figure 8   The Market Adaptation Sequence   80

Figure 9   The Entrepreneurial Dream   100

Preface

Figuring out how to run a business or market a new product in a fast-changing industry is quite a challenge. Putting the ideas into a book is, perhaps, an even tougher challenge.

There are two primary problems. First, books are linear. They start on Page 1 and end on the last page. Reality, by contrast, is multidimensional. When I look at the business world, I see a complex web of interconnections. Every issue affects every other issue in complex and sometimes unpredictable ways. There are no simple answers. Reality doesn't fit into a neat list of chapters.

The second problem: Books are frozen in time, while the business world is constantly changing. As the pace of change accelerates, writing business books becomes more difficult. Today's fictions are tomorrow's truths—and today's truths are tomorrow's fictions. Information that is valuable today might be worthless tomorrow. Writing a business book with lasting value is getting tougher all the time.

In short, books have limitations. But so do all other forms of communication. Communicating ideas within the constraints of the medium can be difficult, but it is also an exciting challenge. Certainly, writing this book has proved an exciting challenge for me.

Acknowledgments

During my twenty years in Silicon Valley, I have worked with more than 150 companies. Many people have contributed to my ideas about marketing and business. But three individuals have been especially important to me, providing me with special friendship and guidance. They are Don Valentine, Andy Grove, and Steve Jobs.

I worked for Don Valentine at National Semiconductor from 1967 to 1970 during his reign as director of marketing. Don now heads a successful venture-capital firm. He has been one of the most influential marketing people in high-technology business since the early days of Silicon Valley, and his influence continues with many startups today.

Andy Grove is the president of Intel. For the last fifteen years, Intel has been the most important force in the microelectronics revolution. It commercialized the first semiconductor MOS memory, the first microprocessor, the first programmable memory, and many other firsts. Andy has influenced not only me, but everyone associated with Intel. His relentless drive, discipline, and imagination are infectious.

Steve Jobs is a friend and a unique businessman. Steve makes people rethink their prejudices. In Steve's mind, nothing is insurmountable. To many people, that might seem foolhardy. To most people associated with Apple, it is inspiring.

I also have been greatly influenced by the writings of two individuals: author/futurist Alvin Toffler and Harvard Business School professor Theodore Levitt. I have read Toffler's *Future Shock* over and over, and I consider anything by Ted Levitt worth reading.

The many people at Regis McKenna, Inc., also deserve a great deal of credit. They are responsible for much of the work and success of the company and receive too little of the credit.

Thanks also to Ann Dilworth and Doe Coover, my editors at Addison-Wesley.

A young man named Ron Davis, while an MBA student at Stanford, helped me research and compile the basic material for the book.

Mitchel Resnick, formerly a reporter and writer for *Business Week* magazine, turned my basic material and thoughts into readable prose. Without Mitch's help, this book certainly would not have come to fruition. We share its creation.

Today change is so swift and relentless in the techno-societies that yesterday's truths suddenly become today's fictions, and the most highly skilled and intelligent members of society admit difficulty in keeping up with the deluge of new knowledge— even in extremely narrow fields.

Alvin Toffler, *Future Shock*

1 The New Marketing

For the past twenty years, I have been going to the best business school in the world. It's called Silicon Valley.

When most people think of Silicon Valley, they think of advanced technology. Bits and bytes, chips and RAMs. But for me, the excitement and importance of Silicon Valley extends beyond the new technologies. For me, Silicon Valley has served as a laboratory for new ideas about business and marketing.

When I moved from Pennsylvania to California in 1963, Silicon Valley did not even have its name yet. But the region was already something special. Collections of engineers were experimenting with new technologies and creating new industries. The pace of change was extraordinary. Every day, it seemed, there were new innovations, new products, new companies.

Though technology was racing ahead, the ways of doing business in Silicon Valley had not changed very much. Companies were still marketing their products in very traditional ways. It was engineers selling to engineers. Salesmen stressed technical details and prices. In short, the businesses were technology- and sales-oriented.

Before long, it became clear to me these traditional approaches were not well suited to a world of fast-changing markets and complex products. Companies in Silicon Valley needed marketing strategies as new and innovative as their products. So while the engineers of Silicon Valley were experimenting with new technologies, I began experimenting with new ideas in marketing and communications.

In 1970, I started my own company, Regis McKenna, Inc., to try out some of these ideas. Through the years, my colleagues and I have worked together with some of the most innovative and dynamic companies in American business. For more than a decade, we have served as consultants to top management at Intel, the innovation king of the semiconductor industry. And we began working with Apple Computer when founder Steve Jobs was still working in his garage and had less than $1,000 in his bank account. In all, we have worked with more than 150 high-technology companies.

Through these experiences, we have helped develop a new approach to marketing, an approach that takes into account the dynamic changes in industries and markets. It is an approach that stresses the building of relationships rather than the promotion of products, the communication of concepts rather than the dispersal of information, and the creation of new markets rather than the sharing of old ones.

While these marketing ideas were conceived and tested at high-technology companies, they can be applied to many other industries. Indeed, the traditional rules of marketing are breaking down in a growing number of industries. The business world is changing quickly, and marketing ideas are lagging behind.

In many ways, the experiences of Silicon Valley can act as a valuable guide to the rest of U.S. industry. Silicon Valley, sitting on the leading edge of change, provides a way for American industry to look into the future. The marketing and management problems Silicon Valley companies faced yesterday, the rest of the business world is facing today. And the problems Silicon Valley companies are facing today, other companies will face tomorrow.

There are several key forces underlying the need for new approaches to marketing. Most important is the quickening pace of change. Traditional marketing rules were invented for static markets and static industries. They assume technologies and markets only change slowly. But the modern business world is anything but static. Advances in technology are causing products and companies to change more rapidly than ever before. Business strategies that look promising one day look obsolete the next.

Technological change used to come along gradually. It took fully thirty years for antibiotics to go from research concept to commercial reality. It also took thirty years to commercialize the zipper. Things are much different today. Products based on recombinant DNA reached the market less than a decade after the first genes were spliced.

Alvin Toffler was one of the first to note the quickening pace of change. In his book *Future Shock*, he explains:

W hether we examine distances traveled, altitudes reached, minerals mined, or explosive power harnessed, the same accelerative trend is obvious. The pattern is absolutely clear and unmistakable. Millenia or centuries go by, and then, in our own times, a sudden bursting of the limits, a fantastic spurt forward.

This speed-up is most extreme in high-technology industries. Just as one year in a dog's life is equivalent to seven years in a human life, one year in the high-technology business is like seven years in any other industry. According to one estimate, there is an innovation every thirty seconds in Silicon Valley. Products move from drawing board to marketplace with lightning speed. Today's new idea is tomorrow's new product.

The pace of change is now speeding up in old-line industries as well, causing major headaches for old-line managers. The same technologies that helped create Silicon Valley are invading more mature industries. Microprocessors, computers, and robots are proliferating, changing the nature of business in almost all industries. Every new semiconductor chip, every new computer, every new software package can profoundly affect the direction and pace of change in other industries.

In the past century, oil served as a base industry, creating new opportunities and changes in industries ranging from transportation to chemicals and textiles. Microelectronics is now emerging as a new base industry. Many industries, like appliances and automobiles, are designing microelectronic devices into their products. Other industries are using microelectronics to automate their production processes. In all cases, microelectronic devices are accelerating the pace of change.

By the end of the decade, no industry will be left unaffected. Even an industry like the clothing industry, hardly a new business, will undergo major changes. Designers will do their work on computer terminals rather than paper. When they are finished with the design for a new dress, they will push a button and send the design to a computer on the factory floor. There, ultra-precise lasers, controlled by computer, will cut the material for the dress. Other machines will sew the dress together.

New styles and fashions will come to the market more quickly than ever before. The day after a designer finishes his or her work, the new fashion could be on the retailers' racks. Electronic signals from the designer's computer will give new instructions to the factory computer, and the laser will do the rest.

This type of system will give clothing manufacturers incredible new flexibility. Traditionally, most clothes have been produced in a limited assortment of styles and sizes. That will no longer be necessary. To change the size or style of a dress, for instance, manufacturers will simply reprogram the laser. The machine that produces a size 6 also will be able to produce a size 16.

Already, we are shifting out of an age of mass-manufactured goods and into an era of custom-made products. Unlike the black automobiles from Henry Ford's assembly line, the new products come in different shapes, sizes, colors, and varieties. Today, diversity costs no more than uniformity.

In this new environment, marketing managers must learn to treat every customer as an individual. As time goes on, customers will demand more and more diversity in their products. Each person wants to be a little different, a little special. Why did so many people want Cabbage Patch dolls? Because each doll was unique, a little different from all the others.

The growing diversity of products is apparent all around us. You can see it in the variety of automobiles on the road, or the great range of clothing styles people wear. The average supermarket in the United States carries about 10,000 individual products or brands. According to a *Wall Street Journal* article, seven new food products are introduced every day. In the first eight months of 1984, for example, no fewer than nine new granola-bar products came to the market.

The diversity of technology-based products is even greater. Dozens of new products are introduced every day. Catalogs for semiconductor companies list thousands of different types of chips. And bookshelves are bending under the weight of today's software catalogs. There are tens of thousands of different software programs available for personal computers. There are programs to manage pig farms and programs to teach young children to read, programs to help baseball coaches make decisions and programs to help movie directors create special effects.

At the same time, products are becoming more complex than ever before. When a customer buys a new computer, he must understand what types of software and peripherals can be used with it. Many customers are confused and bewildered by the choices. It is as if supermarket shoppers had to worry about which types of milk could be used with which types of cereal.

All of these trends—the quickening pace of change, the rise in diversity, the increase in complexity—are creating new challenges in marketing. In many cases, companies are confronted with first-time experiences. That is, they are facing situations that no company has ever encountered before. These situations involve lots of risk and lots of uncertainty.

Clearly, managers are having trouble keeping pace with all of these changes. *Business Week* magazine drove home this point when it reevaluated some of the companies that had been examined—and praised—in the 1982 bestseller *In Search of Excellence*. The *Business Week* article, published in November 1984, found that many of the "excellent" companies had slipped. The primary reason: difficulty adapting to changes in the market.

A 1981 article in *Journal of Marketing* reached similar conclusions. In preparing the article, entitled "Top Management's Concerns About Marketing: Issues for the 1980's," Frederick Webster, Jr., interviewed top executives at twenty-one corporations, including Kodak, General Electric, General Foods, IBM, and Mobil. Almost all agreed that fundamental changes are needed. They argued that marketing managers are "not sufficiently innovative and entrepreneurial in their thinking and decision making."

How do today's marketing managers fail? According to Webster, they fail in several ways:

- They don't provide proper stimulation and guidance for research and development (R&D) and product development.

- They don't exploit and develop markets for new products developed by R&D.

- They don't define new methods for promoting products to customers in the face of major increases in the costs of media advertising and personal selling.

- They don't stick out their necks and take necessary risks.

- They don't innovate in distribution and other areas in order to keep up with the changing requirements of industrial customers doing business on a multinational basis.

- They don't refine and modify the positioning of their products.

In our consulting work at Regis McKenna, Inc., we help companies address these problems. With each client, we examine the company's strengths, expectations, and goals. Then we compare the company's goals with the attitudes, perceptions, and trends of the market. Within this framework, we help the company develop a positioning strategy, a plan to help the company achieve a unique presence in the market.

In most cases, I try to avoid the traditional approaches to marketing. Running more advertisements and mailing out more press releases will not solve today's marketing problems. Customers are already deluged with information. Few people can remember the headline on yesterday's newspaper or the cover on last week's *Time* magazine. In our society, information has become disposable.

Instead, my focus is on understanding the market, moving with it, and forming relationships. While information is fleeting, relationships have a permanence that is very powerful in a fast-changing world. Managers might not remember yesterday's headline, but they will remember the people they had lunch with last month and what each person said. By forming the right relationships, a company can gain credibility and recognition that it would never gain through advertising.

At the same time, I urge companies to view marketing as an educational process. The complexity and diversity of today's products confuses and intimidates many customers. When customers are confused, companies must find ways to educate them. When customers are intimidated, companies must find ways to reassure them. Simply boasting to customers about the speed or power of the product will not do the job.

Most broadly, managers need a new way of thinking about marketing. They must be creative, smart, aggressive, and open to change. They cannot be locked into the ways of the past.

Many industries are entering uncharted territory, so case studies are of little use. Managers should try new things, new approaches.

All the statistics and analyses in the world cannot predict what will happen next in these technology-rich times. Managers need to learn to live with this uncertainty. They must gain a qualitative sense of perceptions and trends in the market. They have to be willing to take some risks, and cut their losses if things go wrong. They need to keep in touch with their customers, and develop an intuitive knowledge of the marketplace. Intuition will carry them a lot farther than a book full of statistics.

Above all, managers must be willing to modify their plans as the market changes. Companies, technologies, and products are always changing, so marketing strategies must change as well. Marketing a product is a continuous experiment. Nothing is certain. Managers must monitor and modify, monitor and modify.

In this book, I'll expand on these ideas, explaining how managers can establish strong positions for their products in these rapidly changing times. The book's focus is clearly on marketing, but its message is important to all managers. For more than ever before, marketing has become intertwined with other aspects of business. Marketing issues affect all other parts of a business, and, conversely, all other parts of a business affect the development and implementation of marketing programs.

Most companies compartmentalize business functions, keeping marketing activities separate from finance, R&D, and other functional areas. In many cases, people in separate areas don't even talk to one another. But that separation is artificial. To succeed in the new business environment, all managers must think about marketing.

Product designers, for instance, should think about marketing relationships with other companies. Some new products make use of a dozen or more different technologies. No one company can stay at the forefront of all the necessary technologies. Marketing relationships are essential to the product-development effort.

Similarly, finance and marketing are intertwined. When customers buy complex and expensive products, they worry about future support for the products. If a company reports weak financial results, potential customers could be scared away. On the other hand, strong financial results will pull in even more customers.

In the next seven chapters, I hope to show how companies can apply new ideas about marketing throughout their organizations. In this framework, marketing is more than a compartmentalized activity. It is a new way of thinking that permeates the entire company. These new ideas are important for all managers who must cope with change. And in today's business environment, that means just about everybody.

You see things and you say why, but I dream of things that never were and say why not.

George Bernard Shaw

2 Dynamic Positioning: The Cornerstones of the New Marketing

Thinking Dynamically

If you say "instant photography" to someone, they'll probably think of Polaroid. If you say "innovative computers," chances are they'll think of Apple. If you say "high-quality copying machines," they might think of Xerox.

Each of these companies has succeeded in positioning itself in the marketplace. Each has established a unique presence for itself and its products. That type of presence is a powerful force in marketing. Indeed, at the heart of every good marketing strategy is a good positioning strategy. Modern marketing is, to a large extent, a battle for positioning. If companies are to develop a new style of marketing suited for the new era of rapid change, they must start with a new approach to positioning.

Positioning is always competitive. Customers think about products and companies in relation to other products and companies. They set up a hierarchy in their minds, then use that hierarchy when making decisions. When people think about copying machines, they put Xerox at the top of the hierarchy, followed by IBM and perhaps Ricoh. In airplanes, people put Boeing at the top of the hierarchy. In military electronics, it's

Lockheed. Once a position is established—be it negative or positive, leading or following—it is extremely difficult to change.

Positioning is vital to success in marketing. All of marketing—merchandising, advertising, pricing, packaging, distribution, public relations—grows out of positioning. If a company's products are positioned poorly, the rest of the company's marketing strategy will be useless.

How can a company establish a strong position in today's fast-changing markets? It isn't easy. Traditional positioning strategies are not adequate. They do not take technology and change into account. They assume a static marketplace—that is, a marketplace where technologies, products, and customer perceptions change very slowly. For today's markets, marketeers need a new model of positioning. They need what I call "dynamic positioning."

Dynamic positioning strategies are very different from traditional positioning strategies. In the traditional model, a company first decides how it wants its product positioned. The company might want to be perceived as the "low-price" company of the industry. Or it might want to be perceived as the "premium-quality" company. Next, the company comes up with a slogan that summarizes the desired message. Finally, it simply spends money on advertising and other promotions until the slogan achieves broad recognition.

The Avis-Hertz rivalry is a classic example of traditional positioning. Avis decided on a position: the hard-working runnerup in the industry. Then, it came up with a slogan: "We try harder." Finally, it advertised like crazy, until people began to believe that Avis really did try harder. This approach worked because the rental-car business is a rather static market. Neither the cars nor the service changes much from year to year. If you rent a car, you rent a car. Companies can differentiate themselves simply through advertising, discount rates, and free gifts.

The situation is quite different in fast-changing industries. In these industries, products change, markets change, technologies change, competition changes. There is a constant flux. Radical changes occur every few years. New companies, and companies from other industries, are constantly trying to grab a piece of the action. All these changes can influence positioning in the marketplace.

Standard approaches to positioning do not necessarily work. A company that is No. 1 today has no guarantee that it will be No. 1 tomorrow. New technologies can turn a seemingly solid position into a fragile one almost overnight. No amount of advertising can prevent that from happening. Even with the best of slogans, a company can lose its position in the market.

Market awareness is no longer enough. An associate of mine once did an analysis of the Federal Energy Administration's slogan: "Don't be fuelish." About 80 percent of all people surveyed were familiar with the slogan. But consumption of energy was at an all-time high. The slogan was good, but the behavior remained unchanged.

To survive in dynamic marketplaces, companies clearly need a new form of positioning. Companies have to establish strategies that can survive the turbulent changes in the market environment. They must build strong foundations that will not be blown away in the storm.

To do that, companies can't focus on promotions and advertising. They need to gain an understanding of the market structure, then develop strategic relationships with other key companies and people in the market. They must build relationships with suppliers and distributors, investors and customers. Those relationships are more important than low prices, flashy promotions, or even advanced technology. Changes in the market environment can quickly alter prices and technologies, but close relationships can last a lifetime, if not longer.

With this approach, positioning evolves gradually. The positioning of a product or a company is somewhat like a person's personality. Babies have no real personality when they are born, but they gradually gain characteristics as they grow. They are influenced by their parents, then by their friends, then by school. Their personalities alter and grow depending upon the environment that surrounds them.

The situation is similar with positioning. A product or company has no real meaning at first. But it acquires meaning from its environment, and it changes as the environment changes. As a company evolves, it is still the same company, just as a growing child is still the same child. But personality and positioning are always changing.

Unlike traditional positioning, dynamic positioning is a multidimensional process. It involves three interlocking stages—product positioning, market positioning, and corporate positioning. These three stages, which will be covered in detail in the next three chapters, interact with one another in subtle but important ways. Each stage builds on the others and influences the others. Pieced together properly, they create a whole that is much bigger than its parts. But if any one of them is flawed, then the whole positioning process will falter.

In the first stage, product positioning, a company must determine how it would like its product to fit in the market. Should it build a reputation for low cost? High quality? Advanced technology? Should it try to sell the product to all companies? Or just manufacturing companies? Or maybe just certain types of manufacturers? I always advise companies to pay special attention to "intangible" positioning factors, like technology leadership and product quality. Intangible factors are based on customer perceptions, not raw statistics and numbers. Marketing is not a strictly rational process. Low prices and top product specifications do not always win sales. Rather, it is intangible factors that are the keys to gaining strong product positioning.

In the second stage of the positioning process, market positioning, the product must gain recognition in the market. It has to establish credibility with customers. The marketplace must perceive the product as a winner.

To gain a strong market position, a company needs to understand the workings of the industry infrastructure, the network of retailers, distributors, analysts, journalists, and industry "luminaries" who control the flow of information and opinion in the industry. Companies should identify and work closely with key members of the industry infrastructure. I believe that 10 percent of the people in an industry influence the other 90 percent. If a company can win the hearts and minds of the most important 10 percent, its market positioning is assured.

In the final positioning stage, corporate positioning, the company must position not its products but itself. This is done primarily through financial success. If a company is profitable, many of its mistakes are forgiven, if not forgotten. But if the company's profits slip, its image becomes tarnished. Customers are reluctant to buy products, particularly expensive or complex

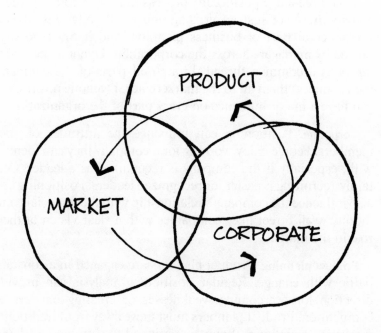

DYNAMIC POSITIONING

Figure 1

products, from a company in financial trouble. When that happens, the company must start over at product positioning and rebuild its position in the market.

This three-stage positioning process must be a total business activity. It is not just advertising and public relations. It is a fundamental part of business planning, and it must be supported by managers across the corporation. Dynamic positioning pulls a common thread through all parts of the company, then connects them all to the marketplace. Dynamic positioning can have a major influence on every part of the organization:

Corporate. Positioning can influence the attitudes of staff members. People enjoy working for a company they can identify with, especially if the company is recognized as a leader. Certainly, recruiting is easier for recognized leaders. Positioning can also influence the company's relationship with the financial community. Wall Street likes companies with a clear vision of their role in the market.

Product planning. Product planners are engaged in a constant battle with change. Regular positioning analysis can provide direction for overcoming weaknesses or creating barriers to competition. Product planners must move away from traditional marketing techniques that were designed to gain market share, and move toward new approaches designed to create entirely new markets.

Marketing. Marketing deals mostly with awareness and the future. Marketing managers must understand the company's positioning, then project the position to the market through education and relationships. Strong positioning allows a company to establish relationships with strong partners. These relationships, in turn, make the company's positioning even stronger.

Sales. Successful sales personnel develop customers, not just orders. Positioning opens doors and provides direction. It gives the salesperson a total picture of where the company is going, and why it is going there. The salesperson is then more confident in conveying this information to customers.

Financial. Positioning and financial strength build on one another. A well-positioned company can raise new funds more easily and at lower rates. Conversely, a financially strong company has a much easier time positioning its products in the market.

The New Cornerstones

At the core of the dynamic positioning process are several key ideas. These ideas differentiate dynamic positioning from traditional approaches to positioning. They influence every stage of the positioning process, and are critical to the development of a successful marketing strategy.

These key ideas serve as the cornerstones of the dynamic positioning process. They provide support and structure to the process. Without them, the process would fall apart.

These new cornerstones of positioning include:

1. Marketing should be dynamic, not static.

Marketing in fast-changing industries is somewhat like guiding a rocket ship from Earth to the Moon—without any sophisticated navigational equipment. In both cases, the target is always moving. The Moon doesn't stand still, and neither does the market. No two Moon shots are exactly the same. During the flight, you have to keep making adjustments, altering the course. If the rocket (or the product) simply goes in a straight line, it misses the target and fails in its mission.

You can extend this metaphor further (if you're willing to stretch your imagination a bit). Think of Earth as the company selling the product, and the Moon as the market. Just as Earth and the Moon exert gravitational forces on the rocket and influence its course, the company and the market exert their own "gravitational" forces on the product and influence its course.

What are the "gravitational" forces on the product? Look at the *company's* gravitational forces first. These forces include things such as:

Financial resources. Does the company have enough money for the product-development effort? If so, does it also have enough money to support the product with the proper marketing, service, and peripheral products?

Timing. In fast-changing industries, the window of opportunity can close quickly. Will the company bring the product to the market at the right time?

Technology. Does the company have all the technology it needs to develop the product? Is its technology at the cutting edge?

People. In the end, people are the most important ingredient for success. Does the company have top-notch talent in its engineering and managerial ranks?

The company can control all these forces, at least to some degree. In some cases, these forces hold the product down. In other cases, they help give the product a strong liftoff.

The *market's* gravitational forces influence the product at the other end of its journey. They draw the product in, help position it in the minds of the customers. They help give a product credibility—or rob it of credibility. These forces include things such as:

Market infrastructure. The infrastructure includes everybody that can influence perceptions of the product: retailers, distributors, financial analysts, manufacturers of peripherals. Support from the infrastructure is critical to success.

Strategic relationships. Companies can form all types of relationships—equity investments, joint development ventures, marketing agreements. A company's credibility in a market often depends on the relationships it forms. For example, Microsoft's credibility in the software industry shot up sharply when IBM decided to use an operating system from Microsoft for its personal computer.

FUD. This stands for Fear, Uncertainty, and Doubt. If customers have fears and doubts about a product, the product won't sell well, no matter how technologically advanced it is.

Adaptation sequence. The market adapts to new technology in stages. First a handful of future-oriented customers (the "Innovators") will try a new technology. Then come the majority of the customers. Finally, the "Laggards" adapt to the technology. Where a product falls in this adaptation sequence certainly influences its chances of success. (See Figure 8.)

Competition. The actions of competitors can turn a product into a smash—or a flop. A product might look good at the launch, but a new product, using a new technology, can make it look obsolete overnight.

Social trends. The prevailing views of society can greatly influence how a product performs in the market. The growing interest in environmental matters, for instance, gave a big boost to solar-energy products.

These gravitational forces are always shifting and changing. Nothing in the market is static. Marketeers will succeed only if they constantly evaluate the gravitational forces and react to changes in the forces. Competing in a dynamic market requires a dynamic marketing strategy.

2. Marketing should focus on market creation, not market sharing.

Most people in marketing have what I call a "market-share mentality." They identify established markets, then try to figure a way to get a piece of the market. They develop advertising strategies and merchandising strategies. All these strategies are aimed at winning market share from other companies in the industry.

In fast-changing industries, however, marketeers need a new approach. Rather than thinking about *sharing* markets, they need to think about *creating* markets. Rather than taking a bigger slice of the pie, they must try to create a bigger pie. Or better yet, they should bake a new pie.

Market-sharing and market-creating strategies require very different sorts of thinking. Market-share strategies are common in mature consumer-goods industries like soft drinks and rental cars. The emphasis is on advertising, promotion, pricing, and distribution. Customers are interested primarily in price and availability. The supplier with the best financial resources is likely to win.

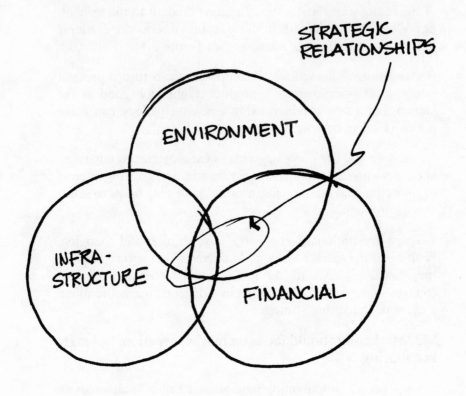

KEY DYNAMIC FACTORS

Figure 2

Market-creating strategies are much different. In these strategies, managers think like entrepreneurs. They are challenged to create new ideas. The emphasis is on applying technology, educating the market, developing the industry infrastructure, and creating new standards. The company with the greatest innovation and creativity is likely to win.

Traditional market-share strategies don't work well in emerging markets. Most new markets are quite small to begin with. If companies think only about sharing the markets, they will never get involved in emerging businesses. They'll take a look at the business, decide that the "pie" is too small, and move on to other possibilities.

That is exactly what happened in the personal-computer business. Dozens of major companies investigated the market for inexpensive computers in the mid-1970s. At the time, these computers were used primarily by hobbyists— that is, enthusiasts who enjoyed tinkering with the machines. There aren't all that many hobbyists in the United States, so most computer companies decided that the market was too small for them to enter.

But a few companies, companies such as Apple and Tandy, looked at the business with a market-creation mentality. They looked beyond the hobbyists and saw that small businessmen and professionals might eventually use the machines—if only the machines were designed and marketed a bit differently. Rather than focusing on what *was*, they focused on what *might be*. They saw the possibility of a growing market, an expanding pie, and they set out to make it happen.

In creating new markets, marketeers face many obstacles. First of all, they can't rely as much on analogies and case studies. When products are radically new and different, past products do not provide much of a guide.

The personal computer, for example, had no good analogies. Clearly, personal computers are not at all like large mainframe computers. They sell at very different price points to very different people. Some people have compared personal computers to stereos. But stereos are much less complicated to use than computers. People don't spend hours learning how to use a stereo. Nor are people scared and intimidated by stereos as many are by computers.

For cases like this, marketeers must break new ground. They must be willing to experiment and take risks. They must try new things and be open to new ideas. Creativity is the key to success in new markets. In mature markets, marketing is like a handball game—a confined environment with few players. But in emerging markets, marketing is more like a soccer game—a wide-open field with lots of players, lots of possibilities, and lots of options.

In the early days of personal computers, no one knew how to distribute the new machines. The traditional method for selling computers—through direct sales forces—was simply too expensive. A person with a market-share mentality might have given up. But a few innovative people persevered. Some tried direct mail, others tried selling computers door-to-door. Still others opened specialty computer stores. Within a few years, there were thousands of computer stores scattered across the country.

To develop new markets, it is essential that companies be willing to take the time to educate customers. When microprocessors were first introduced in the early 1970s, few customers recognized the value of the new chips. People are resistant to change, and the idea of programmable chips was foreign. Many engineers believed the microprocessor was a marketing gimmick.

So Intel, the first company to market microprocessors, had to do a massive education job. It ran advertisements filled with suggested applications for the new product. It distributed booklets with descriptions of actual applications, from electronic games to blood analyzers, from milking machines to satellites.

Most important, Intel ran seminars for potential corporate customers. In the first few years, Intel ran hundreds of these seminars, all over the world. At each seminar, Intel first presented a corporate overview, usually from a top company executive. Next, an Intel marketing manager would give a presentation on the marketing value of microprocessor-based products. Finally, Intel engineers would describe the technical details of the microprocessors. Most of the early customers ordered only a few microprocessor chips. But as the education campaign continued, Intel was able to attract more and more high-volume users.

A market-creating mentality also requires a different view of industry standards. Companies must think about creating new standards rather than following existing standards. That involves greater risk, but the payoffs can be much higher.

Apple took this route when it developed its Macintosh computer. Rather than simply producing a "clone" of the popular IBM personal computer, Apple wanted to develop a new computer that was radically easier to use. To do that, Apple decided to ignore the industry standard in operating-system software, MS-DOS. It developed its own operating system instead. The risks were great, but so were the potential rewards. On balance, it was a risk worth taking.

3. Marketing should be a building process, not a promotional process.

IBM introduced the PCjr in 1983 with a multimillion-dollar promotional campaign. It ran commercials on television and placed advertisements in dozens of magazines. But with all that promotion, the PCjr still didn't sell well. The PCjr's toy-like keyboard was part of the problem. But equally important was the fact that dealers were not excited by the machine. IBM had worried so much about winning the minds of customers that they had never won the hearts of retailers and key industry "luminaries."

In mid-1984, IBM tried to fix the problem. One step was to redesign the keyboard. But equally important, IBM invited all its retailers to a huge meeting in Dallas. Company officials gave the retailers technical information, sales advice, and a big party. They listened to retailers' questions and complaints. The retailers left the meeting with a new attitude toward IBM and a new feeling of commitment to the PCjr. In the next few months, sales of the PCjr began to rise steadily.

The moral of the story? Advertising and promotion are only a small part of marketing strategy. Advertising can reinforce positions in the market, but it can't create positions.

To build lasting positions in the market, companies must first build strong relationships. They must build relationships with suppliers, with distributors, with retailers, and with the financial community. They must take advantage of what I call the industry's infrastructure—that is, the key people and companies that make the industry tick.

I like to draw a distinction between *marketing*-driven approaches and *market*-driven approaches. The two approaches are very different. Marketing-driven approaches are based on advertising and promotion, while market-driven approaches focus on developing strong products, understanding the structure of the market, and building relationships with other people and companies in the marketplace.

Consumer-goods companies usually use a marketing-driven approach. They have a creative idea, then they implement it through advertising. A toothpaste company that wants to increase its share of the market can do so by spending more money on promotions and advertising.

Many technology-based companies have tried to copy this formula. In the past few years, they have rushed to hire executives from consumer-goods companies. Atari lured James Morgan from Philip Morris, Osborne Computer brought in Robert Jaunich from Consolidated Foods, and Apple hired John Sculley from Pepsi.

Other technology companies have begun pouring money into advertising. TeleVideo, hoping to broaden the market for its computers, decided to spend $20 million on advertising. Even Rolm, which sells sophisticated communications systems, decided to launch a national television campaign. According to *Fortune* magazine, computer-industry advertising jumped from forty-fourth place in 1982 to fourth place in 1983, larger than all other categories except food, drugs, and department stores.

Unfortunately, the neat formulas from traditional consumer-products industries do not work well for companies in new fast-changing industries. Rolm has junked its advertising campaign. TeleVideo withdrew from the retail market. Atari and Osborne floundered under their new leaders. Sculley is succeeding at Apple, but only because he focused his initial efforts on product strategy, not advertising and promotion.

Why doesn't the traditional approach work? Largely because customers are scared and confused. They don't understand all the changes occurring in today's fast-changing industries. New technologies seem to emerge every day. Regulatory changes, like the breakup of AT&T, restructure entire industries. Customers in these industries want security and reassurance.

Buying a $1 tube of toothpaste doesn't involve much risk. But buying a $25,000 computer system that will be at the heart of your business is a major risk. Customers are filled with worries: If the computer breaks down, will my business grind to a halt? Will the manufacturer provide prompt and high-quality service? Will my new computer be obsolete in a year? If so, will the manufacturer introduce new, up-to-date models? As my business grows, will the manufacturer offer a smooth up-grade path to larger computers? Will other companies provide the software and peripherals I need for the computer? These are all legitimate fears.

Running more advertisements will not ease these fears. People are deluged with so much product information these days that information has become disposable. With 150 different personal computers on the market, people aren't going to decide which one to buy on the basis of their advertisements. They are going to rely on advice that they get from retailers, consultants, and friends.

There's an old Texas saying about a cowboy who was "all hat and no cattle." That is, he was all show and no substance. Technology-based companies can't build an image that way. They have to have the cattle. If they don't have the cattle, new technological developments will soon leave them in the dust, no matter how strong their advertising and promotion.

For that reason, companies in technology-based businesses must use market- driven, not marketing-driven, approaches. They should concentrate on substance before image, for it is substance that supports the image. They should build relationships with members of the infrastructure who will support and establish their products. Rather than reaching customers through a Madison Avenue campaign, they should reach the customer indirectly, through the retailers, analysts, and other members of the industry infrastructure.

Jim Morgan, the president of Applied Materials, a manufacturer of semiconductor equipment, once noted: "Image is a collection of things that we do in the marketplace." Morgan is not a marketing man, but his instincts are right on the money.

If a company produces a solid product and builds relationships properly, its image will take care of itself.

4. Marketing should be qualitative, not quantitative.

Businessmen love numbers. Numbers make them feel secure. But in emerging markets, numbers are rarely reliable. And marketeers that rely on numbers are unlikely to succeed.

In many cases, quantitative analyses use the past to predict the future. But we live in an era when the future almost never resembles the past. It is extremely difficult to take the pace of technology into account. Extrapolating today's trends into the future almost never works.

Companies have run into this problem since the dawn of the computer age. In the 1940s, computer companies made estimates of the total world market for computers. They calculated the total market at several dozen computers. That's all. Several dozen computers for the whole world. They simply didn't anticipate the proliferation of new applications, or the sharp decline in computer prices.

Mitch Kapor, developer of the incredibly successful software program 1-2-3, ran into a similar problem when he was developing his original business plan for the integrated software package. He developed the business plan for a course at the Massachusetts Institute of Technology's Sloan School in the late 1970s. Kapor received a grade of B for his project, in part because he had no statistical market surveys.

What would Kapor have found if he did a statistical survey? He probably would have found no demand for his product. After all, hardly any large corporations had personal computers in the late 1970s. But Kapor had a sense of the market. He knew that corporations would eventually buy personal computers, and then they would want his software.

Kapor used what I call a *qualitative* approach to the market. He had talked to people in the market. He understood their needs on a human level. A qualitative approach to the market can include many things. It goes beyond the numbers to explore the trends and perceptions that create the numbers. It looks at customer attitudes and personal relationships. Only by understanding the market in a qualitative way can marketeers hope to anticipate the future.

Bare statistics tend to miss the nuances of the market. A survey might show that 60 percent of all customers use a company's product. But a qualitative approach might reveal that the customers are unhappy with the company's service, and many are considering switching to a competitor.

Robert Kennedy once observed, in reference to measurements of the Gross National Product, that we can measure everything except those things which are worth measuring. Yet, as companies grow, they tend to rely more and more on quantitative techniques. They become locked up in numbers and abstractions. They end up with products that do not match the needs of the market nearly as well as the products of entrepreneurs. Creativity is squeezed out of the system.

We need more companies that act like entrepreneurs. The best entrepreneurs don't worry about statistical market projections. They don't care if projections show a $50 million market or a $500 million market. They plan strategy in a qualitative way. They simply take good ideas, develop them into products, then constantly adjust the products to the needs of the market.

A qualitative approach is important in sales as well. Many technology-based companies try to sell their products based on quantitative specifications. They boast that their product has an access time of so many nanoseconds, or a capacity of so many kilobytes.

But customers tend to make their decisions on more qualitative factors, such as service and reliability and reputation. If a company can establish credibility with key people in an industry, it is likely to succeed, even if its product is a few nanoseconds slower than the competition.

To take a qualitative approach to marketing, managers must understand what I call the market "environment." The environment includes all the "gravitational forces" mentioned in the Earth-Moon analogy—things such as social trends, relationships, and competition. Each of these forces (see Figure 3) influences the way in which customers perceive the product.

Quantitative approaches to marketing often ignore the environment. They view products as isolated objects that can be defined by statistics and specifications. But products in the real world are not isolated objects. They exist only in the context of

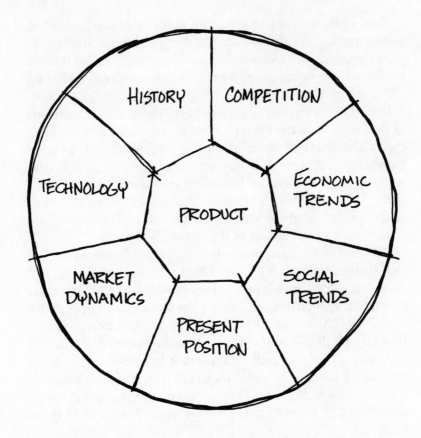

THE ENVIRONMENT DEFINES
THE PRODUCT BUSINESS

Figure 3

their environment. Qualitative approaches to marketing use the environment as a guide to understanding products and markets.

The environment acts as a lens through which the customer views the product. As the environment changes, so does the public perception of the product—even if the product itself has not changed at all. As technology advances, products that were once seen as "cutting-edge" products begin to look mundane. As prices drop, products that once seemed cheap begin to look expensive.

To market a product effectively, marketing managers must understand the workings of the environment. Managers must be sensitive to trends and perceptions. They must understand how various forces in the environment interact with one another, and they must be alert for changes in these forces. In effect, they must see their products as customers see them— through the lens of the environment.

Take the personal-computer industry as an example. Figure 4 shows the market environment that surrounded the 1977 launching of the Apple II computer. The personal-computer industry was still in its infancy, with little competition and lots of future opportunity. Hobbyists were the primary customers, and retail computer stores had just begun to open.

The environment was perfect for Apple. Other industries were full of bad news. Japanese companies were beginning to dominate the automotive and consumer-electronics industries, and they were even making inroads in the semiconductor industry. The American public, and American journalists, were eager for some good business news.

Apple capitalized on this environment. It presented itself as a symbol of hope for the future. It was a bright spot in an otherwise dull and depressing business environment. America was beginning to look upon entrepreneurs as the saviors of American capitalism, so Apple played up the rags-to-riches story of its dynamic founders, Steve Jobs and Steve Wozniak. Time and again, Apple public relations people told the story of the two Steves working late at night in their garage.

Other personal-computer companies emphasized the technical specifications of their products, and made elaborate presentations on the technical differences between brands. But Apple recognized the environment was not a competitive one. The

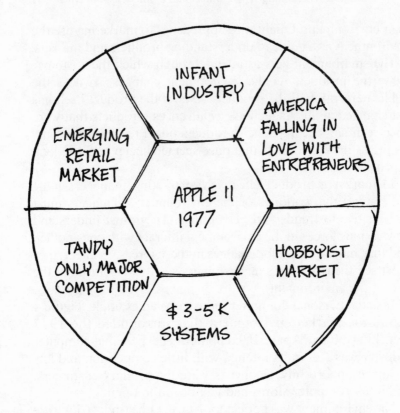

INFANT INDUSTRY

AMERICA FALLING IN LOVE WITH ENTREPRENEURS

EMERGING RETAIL MARKET

APPLE II 1977

TANDY ONLY MAJOR COMPETITION

HOBBYIST MARKET

$3-5K SYSTEMS

1977 PERSONAL COMPUTER ENVIRONMENT

Figure 4

industry was in its infancy and there was room for everybody. Rather, the main challenge was to attract new types of users. So Apple stressed the fun and potential of the new technology.

In short, the 1977 environment for personal computers was nonthreatening and curious. Apple took a qualitative approach to marketing, and it turned the Apple II into a big winner.

The four cornerstones should serve as guides throughout all parts of the positioning process. Once again, they are:

1. Marketing should be dynamic, not static.

2. Marketing should focus on market creation, not market sharing.

3. Marketing should be a building process, not a promotional process.

4. Marketing should be qualitative, not quantitative.

In the next three chapters, I'll examine each of the three stages in the positioning process—product positioning, market positioning, and corporate positioning—showing how companies can put the new cornerstones to use in building a successful marketing strategy.

There is no such thing as a commodity. All goods and services are differentiable.

Theodore Levitt

3 Product Positioning: The Four Golden Rules

The First Step

When Monolithic Memories was still a very young company, its president, Zeev Drori, came to me for some marketing help. He thought Monolithic should start running some corporate advertising to establish a reputation as a reliable supplier. He explained: "We don't want people to think that we just opened our doors." I looked at him and laughed: "But you *have* just opened your doors!"

Like Monolithic, many companies try to put the cart before the horse. They want to gain market recognition and a corporate reputation before they have even introduced a product. But the market doesn't work that way. I always tell companies that the positioning process should begin with the products themselves. Product positioning must be the first step.

The idea behind product positioning is simple. To gain a strong product position, a company must differentiate its product from all other products on the market. The goal is to give the product a unique position in the marketplace.

A company can differentiate its products on the basis of many different factors: technology, price, application, quality, distribution channels, or target audience, to name just a few. A manufacturer of cigarettes, for instance, might try to keep its prices below all other companies in the industry. Or it might target its cigarettes at a special audience, as Virginia Slims did in aiming its cigarettes at women. Or it might try a new distribution channel, perhaps selling cigarettes through the mail.

In some industries, such as the personal-computer software industry, establishing a unique position can be a very difficult task. There are more than 10,000 companies creating and selling software for personal computers. They churn out hundreds of new products each year. Most of the programs get lost in the crowd, and many never even make it onto retailers' shelves.

How can a company gain a strong position in such an industry? It isn't easy, and it's getting more difficult all the time. But it can be done. When I advise companies on product positioning, I stress four key ideas.

First, the company needs to understand market trends and dynamics. I often tell my clients they cannot position their products by themselves. *It is the market that actually positions products.* But if companies understand the workings of the market, they can influence the way in which the market positions their products.

Second, the company should focus on "intangible" positioning factors. Too many companies try to sell their products on the basis of price or technical specifications. It is much more effective to establish positions based on "soft" factors such as quality or technological leadership.

Third, the company should target its product at a specific audience. A company shouldn't try to be all things to all people. It should find a niche. Perhaps it should sell the product only to a certain industry, or maybe it should specialize in a particular application of the product. Whatever niche it chooses, the company should then serve the niche better than anyone else in the market.

Finally, the company must be willing to experiment. With new types of products, no one can be certain of the best positioning ahead of time. A company should experiment with new products, then pay attention to the market reaction. If users suggest changes, the company must shift course and adjust its strategies.

These four ideas make up the Golden Rules of Product Positioning. In this chapter, I'll analyze each of the four rules, examining how each fits into the positioning process.

Understanding the Environment

Imagine two brands of wine. Each is made from the same grapes, stored in the same cellar, bottled in the same type of bottle. Identical in every way. It might seem impossible to differentiate one from the other.

But now imagine that one of the brands is on sale at the supermarket. The other is sold at gourmet food stores and served in fine restaurants. The two are no longer identical. The supermarket brand is perceived as a mediocre wine. The other is seen as a premium wine.

This example shows the power of the market environment. As I explained earlier, it is the environment that "defines" the product. A product cannot be viewed in isolation. The elements of the environment—technology trends, market dynamics, competition, social and economic trends—all influence the way customers "see" the product.

Companies can't just send a positioning message out to the market. They must work with the environment to differentiate and position their products. They must understand what people are thinking, what their prejudices are, what their likes and dislikes are, what they want to hear. Then they must position their products to fit in with the attitudes of the marketplace.

Even if two products have identical features and identical prices, customers might perceive them differently. Maybe the company that produces one of the products has a better reputation for quality. Or perhaps it has better-known investors. Or a more impressive customer list. In any case, the environment makes the seemingly identical products appear quite different from one another.

This is as true for computers as it is for wines. A computer is perceived one way if sold at K mart, another way if sold at Businessland. Sold at K mart, the computer seems like a toy; at Businessland, it seems like a productivity tool.

Timing is also critical, as customer perceptions also change with time. A $3,000 home computer would have been perceived as cheap in 1977, but expensive in 1984. As market conditions change, customer perceptions change—and product definitions change. Product definition is, to a great extent, in the mind of the beholder.

The trick to positioning, then, is to use the market environment effectively. Companies should use the environment to make their products seem unique. Marketing expert Theodore Levitt of the Harvard Business School made this point in his article "Marketing Success Through the Differentiation of Anything." He explained: "Economic conditions, business strategies, customers' wishes, competitive conditions, and much more can determine what sensibly defines the product. One thing is certain. There is no such thing as a commodity—or, at least, from a competitive point of view, there need not be."

Successful companies always consider the environment when trying to position their products. The case of Intel and the "Software Crisis" provides an example. In the early 1980s, Intel noticed its customers were becoming increasingly concerned about the cost and productivity of software development. To make Intel microprocessors useful to a broader audience, companies needed to develop a vast array of software for them. But software development is a slow, expensive, and labor-intensive process. Companies worried they wouldn't have the time, money, or manpower to develop all the new software that was needed.

Intel president Andy Grove understood this environment and coined the term "Software Crisis" to describe it. He explained how Intel products could help ease the crisis. He focused on several new Intel chips. In these chips, some software was actually built into the silicon itself, thus reducing the work for software designers. In this way, Intel products were successfully positioned as solutions to the Software Crisis.

As companies see changes in the environment, they must change the positioning of their products. Consider the case of Measurex. In the early 1970s, Measurex sold a digital computer to paper manufacturers. Using the Measurex computer, paper manufacturers could produce more paper using the same amount of raw material. Customers viewed the product as a productivity-improvement tool, and Measurex reinforced this image through its advertising and sales approach.

But in 1973, the oil embargo caused the environment to shift dramatically. Paper manufacturing is an energy-intensive process, so rising oil prices posed a major threat to industry profitability. In response to this changing environment, Measurex repositioned its product. It ran ads that read: "The Interstate

Paper Company will save 100 barrels of oil a day using the Measurex..."

In essence, Measurex redefined the product as an energy-saving product. Paper manufacturers bought Measurex computers to save energy, and Measurex's sales continued to grow. The product was physically the same as before, but it had a new definition and a new position in the market.

Other companies should work at redefining their products in the same way. In dynamic industries, the environment usually changes from one year to the next. Companies must constantly monitor the environment to spot changes in customer perceptions and attitudes. Then, like Measurex, they can shift their marketing strategies to fit the new environment.

Focusing on the Intangibles

Companies love to make product comparisons. It is common for a company to boast that its product has the lowest price in the industry, or that it is 25 percent more powerful than any competing product. Indeed, an incredible number of positioning strategies center on price and "specsmanship." (That is, promoting a product by its superior technical specifications, or "specs.")

But these approaches to product positioning have serious flaws. Companies are much better off if they establish positions based on what I call "intangible" factors, qualities such as reliability and service. Unlike price and technical specs, intangibles don't fit neatly onto a product-comparison chart. They can't be adequately measured or described by numbers. But intangibles are much more powerful as positioning levers.

Why are intangibles so powerful? First, let's take a look at why price and specsmanship are so ineffective as positioning factors. Competing on price has all sorts of problems. Low-price products are often perceived as low-value products, particularly in consumer markets. Consumers assume cheap in price means cheap in quality. What's more, low-price companies always face the threat that someone else will offer a lower price and steal their position.

The digital-watch business illustrates the point. The first digital watches, introduced in 1972, sold for about $400. But digital-watch manufacturers kept lowering their prices to underprice the competition. Soon, the prices dropped to $99. Then $49. Then $20. Sales of digital watches increased. But hardly anybody was able to make a profit, and many companies left the business.

The idea of the "learning curve," or "experience curve," encourages this type of behavior. A learning-curve strategy involves a two-step logic. First, a company lowers its prices to increase its volume and gain market share. Next, the company takes advantage of economies of scale and mass-production experience to cut its manufacturing costs. Prices are lower, but so are costs.

Unfortunately, the logic breaks down when several companies play the game at once. Prices spiral downward more quickly than expected, and profits follow downward. Makers of semiconductor memories have fought this type of pricing battle several times—to no one's advantage.

Positioning based on specsmanship has similar problems. Companies that position their products as the "fastest" or the "most powerful" often run into trouble. Technological leads are usually short-lived. Research labs develop new technologies every day, and new startups rush to commercialize them. Products move from "leading edge" to "obsolete" more quickly than ever before. As a result, companies that live by specsmanship often die by specsmanship.

There is another problem: Companies that use specsmanship as a positioning lever often ignore the market environment. They see product positioning as an analytic process of product comparisons. They make huge charts showing that Product A can store fifty more kilobytes than Product B. Or perhaps Product A can perform certain tasks five nanoseconds faster than Product B. These comparisons have some value. But they are only the beginning of the positioning process, not the end.

In fact, most customers are not that interested in narrow technical differences between products. Very few people buying personal computers understand the technical differences between one machine and the other 150 on the market. Moreover, they really don't care.

Rather, customers are much more influenced by intangible factors. Intangible factors include things such as technological leadership and product quality, service and support. It's not easy for a company to position a product in terms of intangible factors. The company must build a certain aura around the product. But if it succeeds, it can attract customers and charge premium prices.

The German company Zeiss, for example, has established itself as a leader in optics technology. Its microscopes and telescopes are perceived as the finest available. This image carries over to other products. Some Zeiss sunglasses sell for well over $100, not because of any special features, but because of the Zeiss reputation for quality optics.

The power of intangible positioning became clear to me a few years ago when I was doing a market survey for Intel. As part of the survey, I talked to a number of engineers about a certain memory chip. I remember asking one engineer why he selected the Intel chip. This chip was a fairly technical product, and you might have expected the engineer to answer in technical argot: "The memory had an access time of so many nanoseconds," or "Its power dissipation is only such-and-such."

That didn't happen. Instead, the engineer told me his company buys almost all its chips from Intel, so it was natural to buy the new chip from Intel too. Had he evaluated the new product? Not really. "We just tend to buy from Intel because we have a business relationship there," he explained. "We know where they are going and we trust the company."

I had a similar experience when I visited two small computer manufacturers. Each used Japanese semiconductors in its products. I asked the president of each company why he bought the Japanese chips. The answers were the same: "Quality." I asked if they had done any comparative testing. The answer was no. I asked if they did incoming inspections. Again, no. Yet they were convinced the Japanese chips were higher quality. After all, don't the Japanese have a reputation for high-quality manufacturing? These incidents are hardly unique. Most buying decisions are made the same way. Product managers spend days, if not weeks, drawing up charts and graphs that compare products on the basis of specifications and price. But buying decisions are rarely based on these objective standards. The important product comparisons come from the minds of those in the marketplace. And in people's minds, it is intangible factors that count.

How can a company gain a position based on intangibles? Intel provides a good example. Intel has succeeded in positioning its products as technology leaders in the semiconductor industry. Intel didn't gain this position by specsmanship. In fact, competitors' chips often have superior specs. Rather, Intel's image as a technology leader is based on its people and its production processes. Top executives Bob Noyce, Gordon Moore, and Andy Grove have a long list of engineering accomplishments, and they have often served as spokesmen for the semiconductor industry. Competitors and customers see Noyce, Moore, and Grove as three of the best technical minds in the industry.

Intel has also convinced customers that its proprietary HMOS processing technology is the best in the industry. The process has developed an aura about it. Even when other companies introduce chips that are faster or denser than Intel's chips, Intel is still viewed as the technology leader in the industry. Specs alone cannot dislodge Intel from its intangible positioning.

IBM's products occupy a similar position in the computer industry. Few of IBM products are at the leading edge of technology. Yet IBM is viewed as a technology leader by many customers. IBM buttressed its position in 1982 when it accused two Japanese companies of stealing the designs to its computers. IBM emerged as a defender of American technology. Because the Japanese wanted to steal IBM's designs, people assumed IBM must be the technology leader. IBM's reputation as a technology leader is very solid, and will continue even if there is little evidence to support it.

Reputation works in the other direction as well. Texas Instruments stumbled in the introduction of its first personal computer, and the market won't let TI forget it. No matter how good TI's new computers are, customers still associate TI computers with failure.

Finding the Right Targets

Product positioning is not based solely on the characteristics of the product, be they tangible or intangible. It is also based on how the product is targeted. Companies can build strong product positions by focusing on specific market segments. Baseball old-timers used to say: "Hit 'em where they ain't." Companies can do the same. They can find segments of the market that other companies have ignored, then "hit" into the open spot.

Too many companies try to be all things to all people. They want to become $1 billion companies overnight. I recently met with a small company called Silicon Graphics. The company was less than one year old, but it was already trying to sell into ten different markets. That is a big mistake. A shotgun approach not only taxes limited resources, but also limits the leverage a company might develop by having a significant piece of business in a specific market. It's better to be a big fish in a little pond than a little fish in a big pond.

There are two major reasons why companies should target their marketing efforts. The first reason is obvious. A company that targets its products naturally has less competition. As a result, it has a better chance of establishing itself as the leader in the market segment it chooses. The second reason is less obvious but equally important. When a company focuses its efforts on a particular segment, it can do a better job of understanding and meeting the needs of its customers. And that certainly puts the company in a better position to succeed.

Product positioning is strongest when a company can invent an entirely new market segment as its target. Then, the company can establish a new positioning hierarchy and automatically establish itself as the leader. But in most cases, creating a new segment is not possible. Instead, companies must examine the market environment and decide which existing segments are best suited to their strengths.

Few companies have done as good a job of choosing target markets as Metaphor Computer Systems. Started in 1983 by a group from Xerox, Metaphor developed a computer system that enables managers and analysts to gain quick and easy access to business data. The system gathers statistics from the company's own mainframe computers, combines them with data from outside sources, and organizes all the information in a form that is convenient for the manager to use at his desk.

To market this system, Metaphor recognized that it had to target its efforts. Different types of managers have different information needs, and Metaphor could not hope to satisfy all of them. So Metaphor decided to focus its efforts on particular types of managers in particular industries.

To begin with, the company chose two target industries: consumer packaged goods and financial services. And in each industry, it decided to focus on managers in two functional areas: marketing and finance. Metaphor hired experts from each of the target industries, and developed specialized software packages to meet the needs of the target managers in those industries.

Metaphor also targeted its efforts geographically, initially limiting itself to customers in three cities: New York, Chicago, and San Francisco. The company understood the importance of support and service, and it recogized that it could offer high-quality services only if it limited its geographic reach.

Metaphor's targeting strategies clearly paid off. By late 1984, Metaphor had installed systems at Bank of America, Beatrice, Carnation, and several other companies. The company expects sales to grow to $100 million by the end of 1986.

Once a company finds the right markets to target, it should keep the same focus as it adds followup products. This advice seems so logical, but many companies ignore it. Companies often feel an the urge to expand into new areas where they have little expertise and no established position. Of course, companies must continue to experiment with new ideas. They can't fall into a rut. But they must remember where their positioning strengths lie and take advantage of them.

Digital Research, Inc., is one company that fell into this trap. In the late 1970s, the company became a big success by selling system software for personal computers. Its CP/M operating system emerged as an industry standard and the company's profits soared.

But Digital Research then expanded into "retail" application software—that is, low-end application software aimed at consumers. The retail software business is very different from the system-software business, and Digital Research's culture and expertise were poorly suited for the new business. Its expansion efforts flopped, and the company saw profits drop sharply. Only when the company shifted its focus back to its original target market—system software—did profits rebound.

TeleVideo is another example. In its early days, TeleVideo established a strong position by selling multiuser computer systems—that is, systems used by more than one person at a time. It sold the computers to other companies, which repackaged and resold the systems. It was in a business-to-business business.

Then TeleVideo tried to expand away from its strength. It ignored its established position and tried to sell computers through retail stores. It wanted to build up volume to cut its manufacturing costs. Expenses rose, but volume never followed. Dozens of new computers were competing for retail shelf space, and TeleVideo had no way of distinguishing itself. Retailers were unfamiliar with TeleVideo, and few of them carried the Tele-Video machine. Eventually, TeleVideo was forced to shift back to its strength, selling multiuser computers to middle-man companies.

Experimenting and Changing

Product positioning is not a one-time operation. It is an ongoing process that never ends. A company president once said to me: "How can we ever position ourselves in a marketplace that is changing every three months?" He had a good point. Dynamic positioning is a tricky process. The only way to survive in dynamic marketplaces is to keep the positioning process flexible. Companies must be willing to experiment and learn and change. There is no right and no wrong. In fact, the path to success is often filled with failures.

Marketing people like to think they "know" their market. They do analyses of the market, then develop detailed marketing plans as though the outcome is decided deterministically. But in fast-changing industries, companies are often breaking new ground. No one can really "know" the market. The market doesn't even exist yet.

In these industries, almost all new products are experiments. Few leading-edge products are perfectly in tune with the market when they first come out. Instead, they are modified and altered once they meet the market. There's a lot of give and take.

In some ways, the process is the mirror image of traditional consumer businesses. In traditional businesses, companies survey people to find out what they want, then create a product to fill the need. In technology-based industries, the product usually comes first. Companies invent things and develop things. Then, they work with the market to see how the product should be used.

The process might seem backward, but in some cases it's the only way it can be done. With ground-breaking products, customers can't know what they want until they've seen the product. But after trying the product, they can suggest modifications so the new product or technology fits their needs.

When the first personal computers came to the market, for instance, people didn't know how they might use the new machines. A market research study would have shown very little demand. But a few pioneering companies put personal computers on the market, people came up with suggestions on how to make the machines better and put them to new uses.

The same thing happens with many new semiconductor chips. Semiconductor companies rarely go to customers and ask: "What do you need?" Rather, they take spec sheets to certain key customers and say: "Here's what we can do. How should we modify it to suit your needs?" Successful semiconductor devices often go through ten or twenty revisions during the life of the product. The experimentation never stops.

Beta sites, the locations where a company first tests its product, can be critical in this process. By working with beta-site customers, companies can begin making modifications to the product before taking it to the market. Oftentimes, the beta-site customer will hate the product at first. But as their suggestions are implemented, they fall in love with the product. When the product is finally introduced to the marketplace, it is much more likely to make a good first impression. And that's important, as you never get a second chance to make a first impression.

Experimenting, however, can't end with beta sites. Companies must continue to modify their products and strategies after the product is already on the market. The environment keeps changing, and companies must adapt.

Vitalink is one company that has experimented and adapted well. The company, a well-financed startup in the satellite communications business, originally planned to sell a complete, end-to-end system for sending voice and data messages via satellite. But through market experimentation, Vitalink found the market wasn't really ready yet. Not many companies had a high enough volume of communications traffic to justify the high cost of the service. Even IBM and AT&T were having trouble attracting customers. So Vitalink switched its positioning plan. It began offering specialized services to companies with lower levels of communications traffic. And it began offering some specific applications, like teleconferencing.

Selling into a changing environment can be tough on the nerves. We once did some work for a large company which was launching a product into a new area. The company's managers didn't know how to price the product and kept saying how insecure they felt with this product. But that's only natural. What they were doing had never been done before. You can't be secure unless you have a proven model. In new product areas, there are no models. The answer is to set a price and try it. You have to experiment. The people who usually win are the ones who have the guts to move forward.

All products must be seen as experiments. Many products go through the cycle of failure, change, failure, change. Failure is not necessarily a problem. The critical issue is how quickly a company can react and respond. Managers must first monitor how the market reacts to their product. Then they must modify the product before competitors come up with their own solutions to meet the market's needs.

The stakes are great. Whichever company modifies its product most quickly and most effectively will win the product-positioning battle.

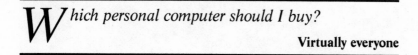

Which personal computer should I buy?

Virtually everyone

4 Market Positioning: Five Ways to Gain Recognition

Gaining Credibility

If you just looked at the numbers, you would think the Apple III computer got off to a pretty good start. The machine, aimed at small businesses, was introduced in 1980. By the end of 1983, the annual sales of Apple III computers topped $100 million. That's not a bad year for a product in a young industry. The Apple III outsold many other computers on the market.

But the market saw the Apple III as a loser. Many people expected the Apple III to be as successful as the company's first major product, the Apple II. The new computer never lived up to those expectations. Because of some early manufacturing glitches and a lack of software, the machine got off to a slow start. It acquired a bad reputation, and it was never able to get rid of its negative image. The product itself didn't have any fundamental flaws. The product positioning was pretty solid. But the Apple III never established a strong market position.

In market positioning, the second phase of the positioning process, the marketplace reacts to the new product. The company finds out whether its product positioning is working. Winning a quick endorsement from the market is critical to success. Once a product wins rave reviews, it picks up momentum in the marketplace. Success builds on itself. The product develops a positive image, and customers flock to it. On the other hand, once the market sticks a product with a "loser" label, the product has a tough time recovering.

Clearly, companies have much less direct control over this stage of the positioning process. Market positioning is determined largely by the perceptions of the marketplace. Customers build a certain image of the product, and no one can argue with their decision.

It is possible, however, to influence the market-positioning process. By understanding the workings of the market, companies can influence the perception of their products. They can create a stronger image for their products. They can take steps to make themselves and their products seem more credible.

Credibility is the key to the whole market-positioning process. With so many new products and new technologies on the market, customers don't know who to believe, they don't know who to trust. Many customers don't even understand the technologies involved in new products. Technology-based products are links in a chain: They are attractive because they are linked to the future. But when people are buying a piece of the future, they need to be reassured. They want to buy from a supplier with credibility.

Quite simply, customers are frightened and confused. To make matters worse, some large companies play on customer insecurities, in an effort to scare customers away from smaller competitors. IBM has turned this fear-raising game into a central element of its strategy. The strategy has become known as FUD: Fear, Uncertainty, and Doubt. IBM salesmen build on the fears that already exist in the marketplace. They portray IBM as the only safe haven in an unpredictable, stormy environment. Their argument can be powerful. Why risk buying from a smaller company? No one has ever lost his or her job for choosing IBM as a vendor.

To establish market positioning, companies in fast-changing industries must find ways to ease customer fears and offset the FUD strategies of corporate giants. They must offset Fear with Comfort, Uncertainty with Stability, Doubt with Confidence. They must build images of credibility, leadership, and quality. They must supply the customer with a "security blanket" in addition to a top-notch product.

How can companies build credibility with customers? Advertising, of course, can play a role. But advertising doesn't get to the heart of the matter. Advertising can only reinforce positions, it can't create them. Increasingly, people are skeptical of what they read or see in advertisements. I often tell clients that advertising has a built-in "discount factor." People are deluged with promotional information, and they are beginning to distrust it. People are more likely to make decisions based on what they hear directly from other people—friends, experts, or even salespeople. These days, more decisions are made at the sales counter than in the living-room armchair.

Advertising, therefore, should be one of the last parts of a marketing strategy, not the first. Companies in technology-based businesses need other ways to build credibility. They must seem secure and trustworthy to customers who are scared by technology. And they must build a solid foundation that will survive the inevitable changes in the market environment.

Of course, companies must start with strong product positioning. Then, they can build credibility—and market positioning—in several ways. The three most important ways are by inference, by reference, and by evidence.

Inference. If a startup is funded by reputable financial backers, or if the startup has a relationship with a respected large company, people infer that the startup must be a credible competitor. Retailers, distributors, and customers begin to take the startup seriously. Compaq and Lotus had instant credibility because Ben Rosen, the well-known venture capitalist, was a lead investor in each company. A deal with IBM also brings instant credibility. Few people had heard of Sytek before IBM chose the small California company to develop a networking product. Now everyone sees Sytek as a technological leader. Similarly, a deal with Burroughs brought credibility to Convergent Technologies, and a deal with Eli Lilly brought credibility to Genentech.

Reference. When people buy complex or expensive products, they often rely on personal references. They'll look for a friend or colleague who has purchased the product, and ask if he is satisfied. Anyone who interacts with the product or the company could act as a reference in the future. Analysts, retailers, journalists, and customers all talk to one another and spread the word about the product. If one person has a good experience with the product, he'll tell others about it, and they in turn will tell still others. Credibility builds and builds. But the process works in reverse as well. A rule of thumb: If a customer has a good experience, he'll tell three other people. If he has a bad experience, he'll tell ten other people.

Evidence. Success in the market reinforces itself. People in the industry look for tangible evidence: rising market share, rising profits, more retailers, new ventures, new alliances. Each piece of evidence adds to the company's credibility and image. As Jim Morgan of Applied Materials said, image is just a collection of the things we do.

There are many things companies can do to put these three credibility-builders to work for them. For example, they can develop relationships with key people in the industry infrastructure. Certain analysts, distributors, retailers, and journalists play a powerful role within an industry. They serve as important references and spread the word about the company. By winning over the infrastructure, a company is 90 percent of the way toward winning the market-positioning battle.

Customers can help a company gain credibility both by reference and by inference. By choosing its customers carefully, rather than simply selling to all takers, a company can control its image. If a small company sells products to a respected corporate giant, such as IBM or General Motors, the small company will probably seem reputable and solid. If it sells to fast-growing startups, it could build an image as an innovative supplier.

Companies can also gain credibility by forming strategic alliances with other companies. Genetic Systems, a biotechnology company based in Seattle, initially seemed similar to many other biotechnology startups: strong in science skills, but weak in business skills. Analysts wondered whether its products could ever succeed in the market. But when Genetic Systems teamed up with Syntex, a large and successful pharmaceutical company, it was suddenly perceived as a legitimate competitor in the medical-diagnostics industry.

Building credibility is a slow and difficult process, but it can be done and it is critical to market success. For the rest of this chapter, I will examine ways in which companies can build credibility, and, in turn, establish market positioning for their products. The strategy can be broken down into five key elements:

- Using word of mouth
- Developing the infrastructure
- Forming strategic relationships
- Selling to the right customers
- Dealing with the press

Using Word of Mouth

A little while ago, I had breakfast with presidents of a half dozen major manufacturing companies. Over orange juice, they discussed the telephone systems their companies were using. Two or three of the presidents were looking to buy new systems at the time. During the conversation, one of the presidents made a remark about a certain company that supplies telephone systems, a company I'll call Company X. "Company X is putting me out of business," he complained. "It seems like the system is always down."

Well, Company X happens to be a very successful supplier of telephone systems, and the company has hundreds of satisfied customers. But that one comment probably made a decisive impression on the other executives at the breakfast. Company X might have the best systems on the market. Its systems might handle more lines than any other competitive product. They might integrate voice and data better than any other system. But the presidents at that breakfast are unlikely to buy a system from Company X. That one offhand comment made them all feel a bit insecure about Company X.

Thus is the power of word of mouth. Forget about market surveys and analyst reports. Word of mouth is probably the most powerful form of communication in the business world. It can either hurt a company's reputation, as in the example above, or give it a boost in the market. Word-of-mouth messages stand out in a person's mind. Memorandums might contain all the correct information, but face-to-face communication is much more likely to gain commitment, support, and understanding.

Quite simply, we find messages more believable and compelling when we hear them directly from other people, particularly when we hear them from people we know and respect. We use word-of-mouth communication to help with all sorts of decisions. We rely on word of mouth to determine what products to buy, what companies to trust, what written reports to read, what corporate leaders to believe.

Information and communication are not the same. Information is objective and cold. Communication is experiential and qualitative. A computer terminal can convey information, but only people can communicate. For communication to be effective, the sender and the receiver must be in sync, on the same wavelength. When people meet face to face, using word of mouth, they are more likely to communicate effectively. Each person can read, analyze, and interpret the attitude of the other person. The same information could be transmitted in a telegram or a written report or an advertisement. But those forms do not utilize the full potential of human communication. They are limited communications tools. Word of mouth turns raw information into effective communication.

Word-of-mouth communication can take on many different forms. Industry participants form "old-boy networks" to keep each other informed about new developments. One recent market-research report showed that such a network plays a key role in the telecommunications industry. Gaining access to the network is critical to success.

Customers use word of mouth too. People are confused about products developed by technology-based industries and they want personal advice. Hardly a day passes without someone asking me that familiar question: "What personal computer should I buy?" Hardly any computer of any size is sold these days without some word-of-mouth reference.

Word of mouth is so obvious a communications medium that most people do not take time to analyze or understand its structure. To many people, it is like the weather. Sure, it is important. But you can't do much about it. You never see a "word-of-mouth communications" section in marketing plans.

As management expert Peter Drucker once noted, more business decisions occur over lunch than at any other time, yet no MBA courses are given on the subject.

Of course, much of the word-of-mouth communication about a company and its products is beyond the company's control. But a company can take steps to put word of mouth to its advantage. It can even organize a "word-of-mouth campaign." Such an effort can be very powerful, because word of mouth is fundamentally different from other forms of communication. It differs in three major ways.

First, it is an experienced process. The message is always carried by a living person. The listener does not only listen to speaker's words. The listener also observes the speaker's tone of voice, facial expressions, hand gestures, and other nonverbal means of communication. The same words could be sent over telephone lines, but they would not carry the same weight. Businessmen recognize this fact, travelling thousands of miles just so they can meet face to face with salespeople and customers. A face-to-face meeting can have greater impact than an avalanche of advertisements, press releases, memos, and brochures.

Second, the message is tuned to the individual listener. The word-of-mouth message can be changed, simplified, altered, embellished, and verified for each person. The message delivered to the director of marketing can be different from the message delivered to the director of engineering. The message can be altered according to the setting as well. A message can be delivered one way in the company cafeteria, another way in a presentation to the board of directors.

Third, feedback is instantaneous. If the listener agrees with the speaker, he will nod or show some other sign of concurrence. If the listener disagrees, he will scowl or suggest alternate arguments. He can help fortify weak points and eliminate irrelevent ones. If the listener does not understand, he will ask for further explanation.

So how can a company harness the power of word of mouth? First the company must decide what message it wants to spread. Word of mouth is most effective in delivering messages about intangible qualities, such as commitment, credibility, appeal, and support.

Then the company must decide who should receive the message—and who from within the company should deliver it. By the nature of word-of-mouth communications, it is not possible to spread the message too widely. Luckily, there is no need to. Word of mouth is governed by the 90-10 rule, which I mentioned in Chapter 2. The rule: "90 percent of the world is influenced by the other 10 percent." So if a company can reach the critical 10 percent, it will indirectly influence all the others. The word-of-mouth message will grow like a snowball rolling downhill, as the critical 10 percent pass the word on to others.

A word-of-mouth campaign should be based on targeted communication. Word of mouth is not an efficient means for distributing information widely. Communications should be directed toward specific audiences. A word-of-mouth campaign aims to develop or change the attitudes and opinions of the people in these target groups, so it is very important to understand something about the minds of the people in the targeted audience. Without this type of understanding, word of mouth is not likely to be effective.

The targets for a word-of-mouth campaign fall into several categories, some of which are discussed in greater detail later in this chapter. The possible targets include:

The financial community. Who backs a company is often more important than how much money is behind it. The community of venture capitalists and private investors is a small, close-knit group. A company's initial backers can use word of mouth to spread the company's message to other venture capitalists, and later to investment bankers, analysts, and brokers.

Industry-watchers. Rapid-growth industries are filled with consultants, interpreters, futurists, and soothsayers who sort out and publish information about the industries. These industry-watchers gain most of their information through word of mouth—visiting companies, attending analysts' meetings, talking to people in the industry.

Customers. Companies can use word of mouth to reach customers at trade shows, technical conferences, training programs, and customer organizations. Beta sites and early customers become especially important. They can spread the word about a new product to other potential customers.

The press. More than 90 percent of the major news stories in the business and technical press come from direct conversations. All journalists have networks of sources they use for background, opinions, and verifications. It is valuable to become part of this word-of-mouth network.

The selling chain. The selling network includes sales representatives, distributors, and retailers. Pricing information and other product data can be distributed through the selling chain in the form of written documents and formal reports. But word of mouth is needed to generate enthusiasm and commitment toward the product.

The community. Every person who is interviewed, or delivers a package, or visits a company walks away with an impression. If company employees communicate properly, every person who comes in contact with the company becomes a salesperson for the company, a carrier of good will about the company.

There are problems with word-of-mouth campaigns. Sometimes it takes a while for the message to spread, and there is always the danger that the message will be garbled as it moves from one person to another. But the benefits of word-of-mouth campaigns greatly outweigh the problems. Companies should learn how word of mouth operates in their industries. Then pass the word.

Developing the Infrastructure

When personal computers first became available in the mid-1970s, most businesspeople saw them as a passing fad. Not Ben Rosen. At the time, Rosen was working as an electronics industry analyst at the Wall Street firm of Morgan Stanley and publishing a newsletter on the electronics business. He began writing, and talking, about personal computers. While others saw personal computers as toys, Rosen viewed them as the basis for a dynamic new industry. Through his newsletter and informal conversations, Rosen began to spread the gospel of personal computing.

Gradually, Rosen began to win converts. First the readers of his newsletter became believers. Then *Forbes* magazine ran an article showing how Rosen used a personal computer for financial analysis. More people started to pay attention to these new machines. Soon, companies began to view a favorable report from Ben Rosen as the key to success in the personal computer business. A bad review from Rosen was the kiss of death.

Why was Ben Rosen so influential? Had he been an obscure broker, his endorsement of the personal computer, or any particular personal computer company, would not have been very significant. But Rosen was considered one of the top analysts in the country. People believed him to be a credible source of information.

Rosen played a key role in the infrastructure of the electronics industry. Every industry has an infrastructure, though it takes a somewhat different form in each case. The infrastructure includes all those people between the manufacturer and the customer who have an influence on the buying process. These people give credibility to the product and the company (or, in Rosen's case, to a whole new industry). Without the support of the infrastructure, the product and company are sure to fail.

I like to picture the infrastructure as an inverted pyramid, with the manufacturer at the bottom and the customers at the top. Figures 5 and 6 show the infrastructures for two industries—personal computers and microprocessors. In each case, information about the product and the company bubbles up to the customer through the infrastructure pyramid, largely through the word-of-mouth process discussed in the previous section.

Each level of the pyramid influences other levels, particularly those above it. Take a look at the personal-computer infrastructure. If third-party hardware and software companies develop products to be used with a new personal computer, industry opinion leaders or "luminaries" begin to take notice. These luminaries, who can be consultants or financial analysts or key users, begin to influence others. Dealers and distributors become interested.

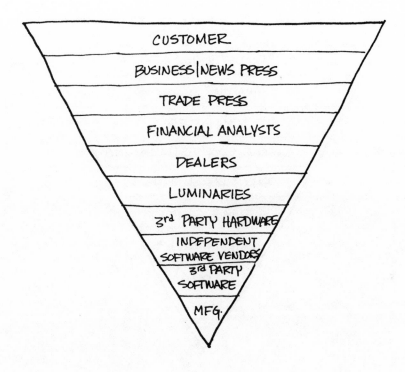

CUSTOMER

BUSINESS/NEWS PRESS

TRADE PRESS

FINANCIAL ANALYSTS

DEALERS

LUMINARIES

3rd PARTY HARDWARE

INDEPENDENT SOFTWARE VENDORS

3rd PARTY SOFTWARE

MFG.

THE INFRASTRUCTURE DEVELOPMENT OF THE PERSONAL COMPUTER INDUSTRY

Figure 5

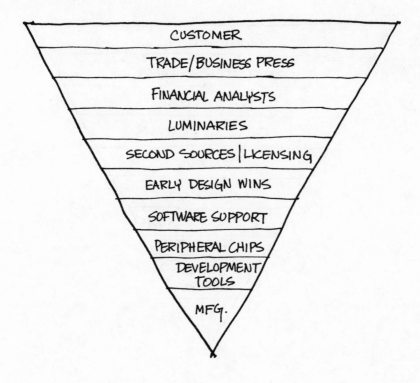

CUSTOMER
TRADE/BUSINESS PRESS
FINANCIAL ANALYSTS
LUMINARIES
SECOND SOURCES | LICENSING
EARLY DESIGN WINS
SOFTWARE SUPPORT
PERIPHERAL CHIPS
DEVELOPMENT TOOLS
MFG.

THE INFRASTRUCTURE DEVELOPMENT
OF THE MICROPROCESSOR
INDUSTRY

Figure 6

Next come the financial analysts. An analyst's favorable report on a new product can significantly improve the product's chances of success. Where do financial analysts get their information? Some comes from the manufacturers themselves. But more comes from dealers and independent suppliers and luminaries.

The business and trade press, in turn, rely heavily on comments and recommendations of the financial analysts. I once went through a long *Business Week* article with a yellow marker and highlighted all the quotes from financial analysts and luminaries. When I was finished, nearly the entire article was yellow.

The spreading enthusiasm bounces up and down the infrastructure. When an influential luminary like Ben Rosen tags a product as a future leader in the marketplace, still more software companies begin writing programs for the computer. Eventually, the word reaches the top of the pyramid and the customers begin buying. It's like a massive game of whispering down the lane, with more and more people involved at each level. Each part of the infrastructure validates the others.

The situation is similar with microprocessors. If a new microprocessor is supported with development tools, peripheral chips, and software, other semiconductor companies are more eager to act as second sources—that is, to make microprocessors based on the same design. This is very important, as few customers want to rely on a chip manufactured by a single company. Second-sourcing agreements attract the attention of industry luminaries and financial analysts, and the word continues to flow up the pyramid toward the customer.

If a company is missing any levels of the infrastructure, the whole pyramid can come tumbling down. National Semiconductor has run into this problem with its 32-bit microprocessor. The product is excellent, superior to some of its better-selling competitors. But the microprocessor has had little in the way of peripheral chips and software support.

As a result, National has not been able to build a reputation as a major force in the market. I recently attended a board meeting at a manufacturing company deciding which microprocessor to use in its next-generation products. Fully three-quarters of the discussion focused on qualitative factors. Board members were looking for a microprocessor maker they could count on for new products and support in the future. National, despite its highly advanced product, was quickly knocked out of consideration. Why? It had no history of success in the microprocessor business and little support from the infrastructure.

The infrastructure tends to be particularly important in rapidly changing industries with complex products. In these industries, there is so much going on that it is difficult for even knowledgeable people to sort out all the details. To understand the significance of new developments, people rely on what they hear from the infrastructure. No dealer is going to open shelf space for a new product without speaking to other dealers or third-party software vendors, or reading trade magazines and financial analysts' reports.

In general, the infrastructure is most critical when the confusion in the marketplace is the greatest. If people are totally secure, they won't ask questions. They will simply buy the best-known brand or the lowest-priced product. But when insecurity is in the air, the company most skilled at lining up the infrastructure will win in the marketplace.

As a result, infrastructure is probably more important for computers than for stereos, and more important for stereos than for cereal. The supporting infrastructure in Silicon Valley is the most sophisticated outside of Wall Street. Analysts, consultants, lawyers, distributors, bankers, and software suppliers all play key roles. Pundits and analysts don't talk nearly as much about new cereals as they do about new computers.

So how can a company line up a supporting infrastructure? I recently talked with a group of junior marketing people at a successful personal-computer company. They told me about their plans to use public relations to create consumer demand for their product. I told them that their plans were bound to fail. Traditional public relations is not enough. Public relations can get your product mentioned in *Time* magazine once a year—if you are lucky. That won't create much demand for a product.

Instead, marketing people must work at identifying and lining up the key members of the infrastructure—and keeping track of how the infrastructure is changing. In the computer business, they must identify the luminaries, the key people in the trade press, the independent software people, then get all of them committed to the new product. They might try starting a newsletter for dealers, or running conferences for industry luminaries. They might give special demonstrations and technical support to independent software companies.

All the time, marketing managers should pay attention to the hierarchies of influence that exist within the infrastructure. Some luminaries are more "luminous" than others. When Intel introduced its first 32-bit microprocessor, it gave an extensive briefing to Gordon Bell, a well-known technology guru, then at Digital Equipment Corporation. When *Fortune* ran an article about the new chip, it ran a quote from Bell to back up Intel's claims about the chip's likelihood of success.

Certain distributors and dealers are more influential than others. K mart is an effective mass distributor, but it will not build credibility for a business computer the way Sears will. So companies would rather that Sears carry the computer, at least initially. Higher up the hierarchy are specialized retail chains such as ComputerLand and Businessland.

In the computer business, lining up the right software companies is even more important. Leading companies like Microsoft and Digital Research can develop "support tools" that make it easier for smaller software companies to develop programs for the computer. And once the software companies commit themselves, manufacturers of add-on hardware, such as plug-in boards and disk drives, are sure to follow.

When a company is able to develop the infrastructure fully, it is almost impossible for its product to fail. The product is a certain success even before it reaches the market. Perhaps the best example is 1-2-3, the integrated software product from Lotus Corporation. Once again, Ben Rosen played an influential role. Rosen, who now heads a venture-capital fund, was the primary investor in Lotus. He began talking about the product months before its introduction. Lots of people had early prototypes. I had one. Many magazine editors had them. We talked to each other about it and the excitement grew. We could hardly wait for the final product to hit the market. By the time of introduction, 1-2-3 was the industry's worst-kept secret, but also its most sure-fire success.

Forming Strategic Relationships

Companies in fast-changing industries need to form all types of relationships. As I have already discussed, they need relationships with venture capitalists, with dealers, with industry luminaries. But just as important, if not more so, are relationships with other companies in the same industry.

In fast-changing industries, these relationships are becoming more important than ever before. As technologies advance and become intertwined with one another, no single company has the full range of skills and expertise needed to bring products to market in a timely and cost-effective way. To produce a personal computer, for instance, a company needs expertise in semiconductor technology, display technology, disk-drive technology, networking technology, keyboard technology, and several others. No company can keep pace in all of these areas by itself.

As a result, collaborative efforts are proliferating. Fast-growing companies, once fiercely independent, are now forming all sorts of alliances, even with former competitors. Every small company, it seems, is looking for "sponsors," while large companies are trying to link up with as many innovative startups as they can. As *Business Week* magazine wrote in a 1984 special report on the computer industry: "For companies large and small, collaboration is the key to survival." These collaborations can take many forms: joint ventures, technology exchanges, manufacturing agreements, and equity positions, among others. Although some of the agreements seem aimed at R&D or finance, these relationships can play a critical role in a company's marketing strategy.

Companies in fast-changing industries need to form strategic relationships for a variety of reasons:

- To compete in today's markets, companies need a diverse set of technologies. Fields like computers and communications are merging, and customers want complete solutions. No company can develop all of the necessary technologies by itself.

- The costs of developing new technologies are rising rapidly. Companies must share the costs if they are to survive.

- U.S. companies are facing increasing competition from Japan.

The Japanese government has led and helped finance cooperative development efforts in fields such as integrated circuits and robotics. U.S. companies must team up to keep pace.

- Technologies are changing more quickly than ever before. At one time, a company could stay at the forefront of many different technologies. Now it is much more difficult.

- Small companies need to gain management expertise, distribution muscle, and capital in order to compete. Strategic relationships can provide these.

- Less tangible, but just as important, strategic relationships can bring added credibility to the companies involved. By choosing the right strategic partner, a company can gain credibility by association.

Many strategic relationships link a small company with a large company. These relationships are not a zero-sum game: Both companies can benefit. Small, growing companies acquire an important aura of credibility by linking up with large, respected companies. The large company acts as a credible reference that tells the market the small company is a winner. Customers are more willing to take a chance with a small company if the small company has IBM or Digital Equipment standing behind it.

At the same time, large companies can gain a window on new technology. Typically, small companies develop new technologies faster than large bureaucratic companies. So by forming links with small companies, large companies can bring more innovative products to the market, and get them there quicker.

A good example of this type of strategic relationship is the alliance between IBM and Microsoft. IBM agreed to use Microsoft's MS-DOS software as the primary operating system on its personal computer. The operating system, essentially the traffic cop controlling activity inside the computer, is a critical element in a computer system. Designers of the operating system and the computer itself must work closely together. For that reason, IBM had always developed its own operating systems for its computers. But the deal with Microsoft made sense for both companies.

For Microsoft, the IBM deal meant instant credibility. Microsoft was an obscure company in Washington state, run by a kid in his 20s. Suddenly, Microsoft was seen as a significant company in the personal computer industry. Its revenues have soared ever since. For IBM, the Microsoft deal meant the giant company could get its personal computer to the marketplace much faster than it could have otherwise. IBM was already somewhat late getting to the market. If it had to develop its own operating system, it might have arrived too late to become a leader.

IBM has forged other alliances as well. To help in the development of floppy disk drives, it struck a deal with Tandon. In microprocessors, it decided to standardize on Intel's family of 16-bit processors. It also invested in Intel, buying 12 percent of Intel's stock, and later increasing its stake to 20 percent. In each case, IBM gained quick access to new technology, while its smaller partner gained an important shot of capital and credibility. IBM's stamp of approval delivered a clear message: This company is a winner.

These strategic relationships allow each company to maintain its independence and unique corporate character. These alliances should not be confused with traditional acquisition and diversification moves. Acquisition strategies often suppress innovation rather than foster it. The larger company often forces the acquired company into its corporate mold, thereby killing the innovative character of the small company that made it attractive in the first place.

Indeed, acquisition strategies in technology-based industries have a pretty dismal record. Schlumberger, for example, tried to acquire its way into high technology. The company, a leader in the oil-services business, wanted to gain a foothold in new technologies, so it began scouring Silicon Valley for acquisition targets. It acquired two Silicon Valley companies: Fairchild, which manufactures semiconductors, and Calma, which develops computer-aided design systems. But the strategy backfired. Key employees left both of the acquired companies, and Schlumberger's corporate culture did not translate well to Silicon Valley. Neither company lived up to Schlumberger's expectations. Schlumberger's desired foothold has turned into nothing more than a toehold, if that.

Other large companies have run into similar problems. Exxon Corporation's effort to enter the office-automation market through acquisition of small high-technology companies turned into a disaster. Western Electric acquired robot maker Unimation in 1982, then saw Unimation's sales drop sharply. And AM International's high-technology acquisitions drove it into bankruptcy in 1982.

Clearly, acquisitions are filled with pitfalls. In many cases, large companies would be wiser to buy minority interests in small companies, or sign development contracts with them. These approaches allow the small companies to maintain their culture and entrepreneurial zeal.

To better understand the growing need for strategic relationships, it is important to understand the product-development cycle in technology-based businesses, as shown in Figure 7. There are many steps between the scientist's workbench and the assembly line, and no company can handle all steps. Strategic relationships are needed to bridge the gaps.

As the illustration shows, the product-creation process breaks down into four stages:

Basic research. Although most industrial advances rely on progress in basic research, industry funds very little of this work. Basic research seeks to answer fundamental scientific questions, such as the internal structure of matter or the properties of human-body proteins. It is a long and uncertain process. No one can know ahead of time whether it will lead to new applications or technologies. Few companies have the resources or patience to fund this type of research. So most basic research in the United States is performed at universities or national laboratories, with funding coming primarily from the government.

Applied research. When a scientific endeavor becomes directed toward a particular industrial result, it becomes applied research. Squeezed between basic research and development, applied research suffers the woes of a middle child: ambiguity and neglect. University researchers, whose main goal is to expand scientific knowledge, prefer to focus on basic research. Small companies cannot afford to get involved until research has already passed through the applied stage. So most applied research is performed at large industrial labs, such as AT&T's Bell Laboratories and Xerox's Palo Alto Research Center.

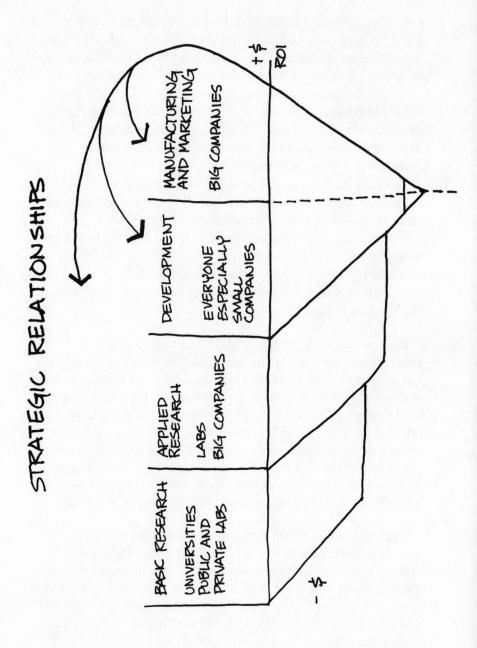

Figure 7

Development. This is the most directed phase of product creation. Its goal is a finished product that can compete in the marketplace. All companies do development work. But in fast-changing industries, small companies are the most productive and successful in development efforts. Unlike corporate giants who must invest substantial resources in maintaining their bureaucratic structures, small startups direct almost all their resources and energies toward development of a product. Necessity presses startups to be more innovative: The company's very survival depends on success of the development effort.

Manufacturing and marketing. The first three stages of the cycle, from basic research through development, all represent investment costs. It is through manufacturing and marketing that companies produce a return and recoup the costs of product creation. Without this return, the investment is lost and it has generated no new capital for the next generation of product innovation. Size and resources are often major advantages in manufacturing and marketing, so big companies tend to be the leaders in this final stage of the product-creation process.

It is in this final stage that Japanese companies have their biggest advantage. Japanese companies usually lag behind their U.S. counterparts in the first three stages of the product-creation cycle. But in some fast-growing markets—such as consumer electronics and semiconductor memories—they have managed to leap ahead in the final stage, in part because of superior manufacturing technologies, and in part because of the special treatment they receive from the Japanese government and Japanese financing system. In winning the manufacturing battle, the Japanese have deprived U.S. companies of the returns they need to invest in future-generation products.

Strategic alliances can help U.S. companies share costs and expertise, and thus meet the Japanese challenge. In looking at the product-creation process, it is clear that different companies have different strengths in different parts of the cycle. Teaming up is a way to share expertise. If big companies are typically stronger in applied research and manufacturing, while small companies are the most innovative at development, why not join forces? IBM's strategic alliances have done just that, linking its manufacturing prowess with the developmental skills of Intel, Microsoft, Tandon, and others.

As competition continues to grow in technology-based industries, strategic relationships will become ever more important. Industry shakeouts are inevitable. Not all 500 computer companies and 10,000 software companies will survive. Small companies have to ask themselves whether they can go it alone without the resources and credibility offered by larger companies. For companies that start to slip, strategic relationships will be the only way to regain credibility, to build a new image. These slipping companies need a dramatic change. Osborne and Victor, two personal-computer companies that went bankrupt, both should have paid more attention to strategic alliances. They didn't and they failed.

In some product categories, such as computer operating systems, one or two dominant standards will emerge in the next few years. For companies developing those products, strategic relationships are particularly important. Those companies must link up with others to help establish their products as standards.

Digital Research, Inc., a leading supplier of operating systems and other personal-computer software, is following this advice. It is now using strategic alliances as a cornerstone of its strategy. The company's CP/M operating system emerged as the industry standard for the early 8-bit microcomputers. But the company stumbled when 16-bit microcomputers, such as the IBM PC, began to replace the less-powerful 8-bit machines. Microsoft, through its relationship with IBM, made its MS-DOS the industry standard in the 16-bit world, and Digital Research saw the world passing it by.

To regain its momentum, Digital Research turned to strategic relationships. Digital Research had more than 200 customers. But none of them were the standard-setters of the industry. That had to change. So beginning in 1983, Digital Research targeted the leaders in the semiconductor, communications, and computer industries. It signed agreements to develop software for Intel and Motorola, the two leading manufacturers of microprocessors. Then, it agreed to develop a library of software applications for AT&T, giving Digital Research a link to the popular UNIX software developed by AT&T. Digital Research also inked a deal with IBM, in which the computer giant agreed to market certain Digital Research software.

Thinking at Digital Research has clearly shifted. With its new strategic relationships, Digital Research has re-established its position in system software. And the company is well on its way to a leadership position in the next generation of system-software products.

Of course, simply establishing the new relationships is not enough. Companies also must know how to capitalize on the links after they are formed. ZyMos is one company that has failed to do that. As a small company manufacturing custom-made semiconductor chips, ZyMos needs to convince customers of its reliability. The marketplace is fearful of a small company in today's environment. All companies are growing so quickly that they need a reliable supplier. They can't afford to have their lines shut down.

Strategic relationships are an ideal way to ease these customer fears, and ZyMos has an impressive list of partners. It has technological agreements with many top-notch companies, including Intel and Apple. But ZyMos is still going out and presenting its product on the basis of technical data. ZyMos's managers came to us and talked for an hour about line widths. That makes no sense. Pitching technology doesn't work in this environment. ZyMos should be stressing its relationships with Intel and Apple. They have the right relationships, but they're not using them.

The idea of strategic relationships is not limited to the electronics and computer businesses. It applies to all fast-changing industries. Strategic alliances can be critical in the biotechnology industry. As in the electronics industry, most of the innovation in the biotechnology industry comes from small firms. But bringing products to market is particularly difficult for small companies in biotechnology. Many biotechnology products must gain government regulatory approval. That is a long and expensive process. Few small companies have the resources to wait out the entire process. Teaming up with large companies solves this problem, while also giving the startups much-needed marketing muscle and credibility.

Strategic relationships helped establish Genentech as a leader in the biotechnology industry. Genentech had a fair amount of credibility from its very beginning because its funding came from one of the most respected venture-capital firms, Kleiner Perkins Caufield & Byers. Using that initial credibility, Genentech was able to attract the interest of Eli Lilly, the pharmaceutical giant. The two companies signed a deal under which Genentech would develop human insulin using recombinant DNA technology and Lilly would produce and market the product. This deal gave Genentech production and marketing capabilities it never could have financed on its own. Equally important, the association with Lilly gave the startup an aura of credibility and established it as the technology leader in the infant industry.

While strategic relationships seem increasingly attractive, there are problems. Nothing is an automatic success. There are many factors working against the formation of strong bonds between companies. Perhaps the biggest source of problems between partners is poor communication. Oftentimes things do not get done because each partner believes the other is responsible. When entering into a new relationship, all companies involved need to be explicit about their objectives and expectations. The companies must agree on all details: What is to be done, by whom, and when. Management responsibilities and financial policies should be clearly stated. In some cases, companies define their markets and goals so differently they always will be in conflict. These types of philosophical differences should be aired and resolved before any agreement is reached.

Antitrust also can be a problem. In light of the economic challenge from Japan and other international competitors, the U.S. government is allowing companies a greater degree of flexibility in structuring alliances. The government has raised no objection to the Microelectronics and Computer Technology Corporation, an alliance of a dozen computer and electronics companies that was formed to share research costs. But other relationships are sure to raise objections. The intent of the partners is the critical factor. Relationships structured to restrict competition are, and should be, unacceptable.

Another problem is that small companies can become overly dependent upon their larger partners. This is similar to the problem faced by military contractors, many of which survive at the whims of the Pentagon. Companies that depend on a single relationship as a primary source of business can end up in big trouble. MiniScribe, a tiny Colorado company that supplies disk drives to IBM, saw its stock plummet by more than one-third when IBM changed its buying patterns. The situation can be even worse when a large company decides to vertically integrate, developing its own production capabilities for parts that it once bought from outside partners. Small companies must remain aware of where they stand in their partner's plans, and should never get in a position where their very survival depends on the continuation of the relationship.

Selling to the Right Customers

Customers are the key to any business. Companies are always looking to attract new customers. However, many companies fail to realize *which* customers they attract is often more important than how many customers they attract.

Just as companies should look to form strategic relationships, they should try to sell to strategic customers. An impressive customer list can give a company a reputation as an innovator or a technological leader. Tandem, the pioneer in "nonstop" computers that never fail, sold its first system to Citibank in New York. To outsiders, the message was clear. If Citibank trusted Tandem, then Tandem must be a winner. *Business Week* quickly ran an article on Tandem, based in part on the Citibank reference. If Tandem had gone to Southwest Mutual as its first customer, it probably would have taken much longer to develop its reputation.

Key customers can help in other ways too. They can give valuable feedback, providing a company with new ideas on how to improve a product. What is more, key customers feed information about the manufacturer into the word-of-mouth network. If every key customer tells two others, and each of them tells two others.... You get the picture.

Companies should pay attention to choosing the right customers even before they introduce their product. Picking the right beta sites to test early versions of a product can be critical to the product's ultimate success. Valid Logic, a new manufacturer of computer-aided engineering systems for the electronics industry, found that one of its beta sites, Convex Computer Corporation, provided important suggestions for improving its product. Convex, itself a high-powered startup, was developing a highly sophisticated supercomputer, and thus was able to push the Valid Logic system to its limits. At first, the Convex engineers complained about the Valid system. But as Valid responded to the criticisms and modified its prototype, Convex became an important Valid supporter.

Convex itself faced an important decision in choosing beta sites for its supercomputer. We had long discussions with Convex about the decision. Should the beta site be at a university? How about a government agency? Or a military contractor? Each possible site had its own unique characteristics. Convex would learn different things from different sites. Equally important, the choice of a site would start to position the company. By choosing a military contractor, Convex might be positioned as a military supplier, and then find it difficult to sell into the commercial market. Ultimately, Convex chose two beta sites: a semiconductor company which is using the computer for chip design and a petroleum company which is using the computer for geophysical research.

Deciding which customers to sell to requires creative segmentation of the market. Many companies selling industrial products give segmentation little thought. They segment the market geographically, or into big and small companies. This is especially true of startups. Ask a startup: What's your market? The answer is predictable: The Fortune 500. They all think the Fortune 500 is some type of magic formula. Sell to the Fortune 500 and you're bound to succeed.

Well, the world doesn't work that way. Fortune 500 companies are large, bureaucratic organizations. They have numerous rules and qualification criteria, and they are generally hesitant to try new technologies and products. Selling to those companies is a long and tedious process, lasting a year or more. Startups would be better off selling to the "emerging Fortune 500"; that is, the 500 companies most likely to grow and be successful in the next decade. These companies have to make purchase decisions more quickly, they are more likely to try new and innovative products, and they probably will come back for repeat purchases as they grow.

Finding the emerging Fortune 500 is not necessarily easy. Tomorrow's winners are not always readily apparent. It's not purely statistical; it's more qualitative. You can't just tear a chart out of *Fortune* magazine. But identifying these companies can pay off. ASK Computer in Los Altos, California, has put this type of strategy to work. ASK sells software to manufacturing companies. Rather than target the Fortune 500, it first went down the street and sold to other high-technology companies in Silicon Valley. Its customer list includes many of the fastest-growing companies in the country. Because these companies do not drag their feet on purchase decisions, ASK was able to quickly grow to a $100 million company.

Another, closely related, approach to segmenting the market involves the "adaptation sequence." Social-science researchers have noted that people fit into four categories according to how quickly they adopt new products and beliefs. Some people lead the way. They are the Innovators. Next come the Early Adopters, then the Majority. Finally, there are the Laggards, who are the slowest to adopt new ideas. According to one book on the subject, about 2.5 percent of the public are Innovators, 13.5 percent are Early Adopters, and 16 percent are Laggards.

These groupings can be used to classify companies as well. Companies have attitudes just as people do, and these attitudes can be used in positioning new products. Only one change: I like to think of companies "adapting" to new technologies, rather than "adopting" them. So I call the process the "adaptation" sequence.

Figure 8 breaks down the personal-computer market into Innovators, Early Adapters, Late Adapters, and Laggards. As you can see, members of different groups have very different motivations and attitudes. Innovators are fascinated with technology and are willing to educate themselves about new products. Laggards, at the other extreme, are much less knowledgeable about new technologies and will not purchase a new type of computer unless there is an absolute need. They respond to competitive pressures. Selling to these different groups requires very different strategies.

The Market Adaptation Sequence

Adaptation Criteria	Small Business/Personal Computer			
	Innovator	Early Adapters	Late Adapters	Laggards
Product Acceptance	Technology Fascination	The Coming Thing	Obvious Solutions To Problem	Absolute Need
Motivation	Implement New Idea	Leap Frog Competition Improve Business	Competitive/Social Pressure Fear of Obsolescence	Extreme Competition/ Social Pressure
Confidence Level	Willing to Experiment High Self Confidence High Risk	Willing to Try New Things Will Go With Reasonable Risk	No Risk Slow to Change Needs References	Reluctant to Change Culture Problems Strong Justification
Education Attitude	Self Taught Independent	Will Attend Night School to Learn	Will Attend Seminar Wants to Buy a Proven Product Needs Lots of Hand Holding	Will Send Someone to Seminar Needs Proof Ease of Use
Acceptance Criteria	Latest Technology New Features Performance	Innovation Better Way to Do Job Selective	Brand Important Pay For Only Needed Features Terms and Conditions Important	Lowest Cost Competitive Terms And Conditions Brand Very Important
Selling Strategy	Self-Sold, Once Turned on Word of Mouth	Benefits Reference Word of Mouth	Address Cost Problems/ Technical Support Needs Examples Demonstration	Productivity Increases Fear

Figure 8

There is a great temptation to target Laggards in your marketing strategy. First of all, there are generally many more Laggards in the marketplace than Innovators and Early Adapters. What is more, Laggards are typically large corporations that could lend immediate credibility to your business.

But it usually makes more sense to aim new products at the Innovators and Early Adapters. Innovators are more likely to take a chance with a new product or new technology. And because innovators are usually, though not always, small companies, they make purchase decisions more quickly. Moreover, the actions of Innovators influence all the others. Innovators spread information about the product through word of mouth. If Innovators buy a product, others are likely to follow suit. So selling to an Innovator might actually bring more credibility than selling to a Laggard, even if the Laggard is many times larger and better known. As Innovators influence companies downstream in the adaptation sequence, credibility for the product grows and grows.

A comment from a senior vice president at General Electric backs up this idea. GE is a major Intel customer, and I would classify it as a Late Adapter. GE made a commitment to use Intel's 8086 microprocessor in a variety of its products. But when I talked to a senior manager at GE in 1982, he was getting nervous about the decision. He noted that many innovative companies in Silicon Valley, including Apple and Fortune Systems, had decided to go with Motorola's 68000 microprocessor. Intel's 8086 still had dominant market share, and its customer list included many major companies. But the Motorola chip took on a certain aura. As the GE executive said to me: "Those startup companies in Silicon Valley are capturing our imagination. If we had to make the decision today, we would not go with the 8086."

Large companies can sometimes afford to wait until Laggards begin to buy a product. Then, they can sell in large volumes to the Laggards. But small companies cannot afford to wait for the Laggards to come around. They must target the Innovators, preferably the most visible of the Innovators. If I were a startup selling disk drives, and I could choose to sell to Apple, Xerox, AT&T, or a startup like Metaphor, I would probably choose Apple and Metaphor. The order might be for fewer drives, but my reputation would be established. Rather than spending millions of dollars on promotion, I could simply rely on word of mouth.

Of course, identifying the Innovators in an industry is not always easy. Not all Innovators are small. Some pockets of large organizations are Innovators. At banking giant Citicorp, for example, the MIS division is an Innovator. At Xerox, some divisions are Innovators, while others are Laggards. So you must understand the organizational forces acting within large companies to decide whether or not to target certain divisions. Intel has an Innovators Program that identifies which divisions at other companies are Innovators, and which are not. Other companies might benefit from similar programs.

Dealing with the Press

I put this section last because it should come last. Too many companies think press relations come first. They want to make a splash in the press even before they position their products. They think that a good article in *Business Week* or *Fortune* or *The Wall Street Journal* can create their markets and solve their problems. They believe a strong media campaign can make up for deficiencies in product quality, customer relations, and other basic marketing skills.

These ideas are totally backward. Press relations cannot change reality. Press relations do not create what you are; they reflect what you are. Press relations cannot take the place of a broad-based marketing strategy. Companies must first position themselves and establish themselves in the marketplace. Then, and only then, should they worry about getting press coverage.

Now that I have deflated the importance of press relations, I should emphasize the flip side: When handled properly, press relations can be a valuable part of a company's marketing strategy. Indeed, a company is unlikely to succeed without good press relations. A company can lose in the press and still win in the market in the short term, but that can't happen in the long term.

Press relations do not have to be all "fluff." They are serious business. Once a product is positioned, press coverage can help reinforce and broaden the credibility that the product and company have already gained. The press can ease customer fears and make customers feel more secure about new technologies. In new and fast-growing industries, journalists can play the role of evangelists. They can preach the new technology.

Advertising can perform many of the same functions. But press relations is usually more effective and credible. Articles in the media are perceived as more objective than advertisements. If a company can win favorable press coverage, its message is more likely to be absorbed and believed.

Press relations serve a second purpose: They can provide a company with valuable feedback. Communications is a two-way street. Companies can learn a great deal from journalists. Like analysts and other industry observers, journalists serve as a microcosm of the world at large. By talking to the right journalists, a company can learn much about how the world views its products and the company itself. This type of feedback can be invaluable as a company attempts to fine-tune its public image.

If a company has monitored the environment and thought about positioning, it should have no difficulty figuring out what message to deliver to the press. The message should evolve naturally from the positioning process. But it is not always easy to deliver that message in an effective way. Communication seems so simple. Yet, so few companies do it well.

Successful press relations requires time, planning, and constant reinforcement. It rests with an understanding of how journalists work and how information is communicated. I have put together a set of guidelines that can be useful in developing an effective public-relations strategy:

Understand the journalist's role. Journalists value their role as independent observers. They resent companies that try to blatantly influence them or co-opt them. They do not want to be viewed as an extension of the company's promotional efforts. The notion that a story is free advertising is degrading to the journalist and to journalism.

Companies must present information without trying to manipulate. Manipulation can be counterproductive. Litronix, which sold light-emitting diodes in the early 1970s, learned this lesson the hard way. The company saw its sales starting to turn downward, and it decided an article in *Business Week* would help revive the business. Sure enough, *Business Week* was interested in an article. But the headline read: "Burnout of a Star."

Don't go to the press too early. Obviously, no one wants negative press coverage before a product is even introduced. But positive coverage can be almost as bad. A favorable article while the product is still in development might build expectations that are difficult to meet. If problems crop up and slow the development cycle, as so often happens, the whole world will know.

Synapse Computer ran into this problem. Synapse had impressive credentials. Started by a group of engineers from Data General, the company planned to build "fault-tolerant" computers that would never break down. They had an excellent chance to succeed in the market. But they had the itch to tell the world they were great before they actually were. Even while still working in the back of a candle factory, they began running ads and talking to the press. Expectations rose. Then, Synapse's computers ran into technical problems at the beta sites. Nothing that abnormal, just typical beta-site problems. But Synapse was very visible now. Journalists were watching Synapse, and they reported on the company's problems. Synapse's credibility sank like a rock. Whether or not it solved the technical problems, Synapse faced an uphill struggle.

Don't "imprint" the wrong image. When a baby chick is born, it looks around for its mother. If the first thing it sees is a human, it assumes the human is its mother, and its mind will never be changed. This process is called "imprinting." Customers often act the same way. When a startup company introduces its first product, customers will form an image of the company, and that image is very hard to change. In short: You never have a second chance to make a first impression.

3Com, a small company that develops communication networks for computers, managed to avoid this problem by being patient. The company developed the first personal-computer network compatible with Ethernet, the industry standard for larger computers. But 3Com faced a promotion dilemma. Not all pieces of the network were ready at the same time. It wanted to introduce each piece—the software, the controller, the transceiver—as it was ready, so it could start receiving revenues and gain market experience. But the company didn't want to be perceived as a component company. It wanted to be seen as a full-systems supplier. With our advice, 3Com waited until all pieces were in place, then began communicating its message to analysts and journalists. The strategy worked: The company is now firmly positioned as a systems supplier.

Get the infrastructure ready. Most journalists practice what I call "he said, you said" journalism. Rather than present their own analysis, they simply quote what other people say. And who is it they quote? Most often, they quote members of the industry infrastructure—financial analysts, consultants, distributors, early customers. The infrastructure serves as a type of filtering mechanism, helping journalists separate fact from fiction.

Companies should take advantage of this filtering mechanism. They should educate and win over members of the infrastructure before going to the press. If a company tries to go to *The New York Times* or *The Wall Street Journal* without first developing the infrastructure, it could run into big problems. Reporters will go to members of the infrastructure, and the company might not be happy with what the people in the infrastructure say. Clearly, what people in the infrastructure say about you matters a great deal.

Meet with journalists one on one. Many companies build their press strategies around press releases and press conferences. But these are not the most effective ways to communicate a message. National magazines get thousands of press releases every week. It's tough to get heard through all the noise. Many press releases are thrown out without being read.

Nor are press conferences very effective. There are two problems. First, journalists are reluctant to ask their best questions at a press conference, because they don't want to tip off the competition. Second, different parts of the media have different interests. *Byte* magazine wants to hear about nanoseconds and megaflops. The newsmagazines want the broad trends and social implications. It's impossible to satisfy everybody. There's a lot of information, but not much good communication. A press conference is a nice spectacle, but the press loses out—and so does the company.

Instead, companies should meet members of the press individually. A one-on-one meeting takes more time, but it makes more of an impression on the journalist and it delivers the message more efficiently. Messages can be tailored for the audience: one for the trade magazines, another for the business magazines, a third for the general-interest press. Once again, the 90/10 rule applies: 10 percent of the press influences the other 90 percent. So select the most influential members of the press and meet with them.

Educate the media. Press relations should be seen as an education process. Fast-moving industries are becoming more diverse, fragmented, complex, and difficult to understand. At the same time, there is more information available about every facet of every industry. For most journalists, these industries are becoming more and more confusing.

Companies need to help journalists create order out of the chaos, so journalists can present a cogent description of emerging trends and technologies. Rather than simply pitching ideas to the press, public relations people must be willing to spend time and educate the press. Companies should treat journalists as well as they treat their major customers. It's not enough to hold a new product up to a press conference of 600 people and say: "Here it is."

Develop long-term relationships. Developing good relationships with the press takes time. Press relations is a process not an event. Pressing the media for an immediate article will rarely succeed. Most major business stories take months, even years, to evolve. Companies must be patient.

Companies should view press relations as a continuing investment. It will pay off with time. Once you establish good relationships with the media, you will be able to present new products more effectively. Moreover, you will be able to participate in broader articles about industry trends, and you will become less susceptible to speculative stories. Journalists will seek your side of the story before going to press.

Look beyond products. In new industries, the press typically focuses on products. The stories are generally naive and superficial. Most of the coverage comes from the trade press. But as an industry matures, so does press coverage. Journalists learn, question, dig into the "news behind the news." The business and general-news media become increasingly interested.

Companies must deal with the business and general-news media differently than they deal with the trade press. There should be much less emphasis on product performance and characteristics. Seasoned journalists know a technological advantage is short-lived. Companies should explain how they fit in the present and future business environments. When products are discussed, they should be placed in a broader context, such as "The Office of the Future" or "The Factory of the Future." The press is fascinated by glimpses of what lies ahead.

Be honest about bad news. When bad news strikes, it's not worth fighting the press over it. As a politician once told me: "Never pick a fight with someone who buys his ink by the barrel." Being honest scores points with the press. In negative situations, a company's character and style will greatly influence how the press perceives and writes about the company.

It almost never makes sense to hide bad news. It is best to get the bad news out, so it's over and done with. If you try to hide the news, it will fester and go on forever. Three Mile Island is a classic example. The Nuclear Regulatory Commission withheld information, and public confidence sank lower and lower. On the other hand, Johnson & Johnson was very open with journalists during the Tylenol scare, and Tylenol has since regained its credibility in the market.

Use top management. At most companies, top management pays attention to press coverage only when the coverage is negative or when the competition receives a lot of positive coverage. I believe top managers should play a more active role in press relations—and marketing in general.

At most companies, especially small, technology-based companies, the personality and culture of the company can be traced to the management team. As the company grows, and marketing plans proliferate, that corporate personality often fades. Top managers are then the only ones able to communicate the corporate character and ideals. They are the only ones who can offer a simple, unified view of the total corporation.

If you put layers of people between company management and the journalist, the journalist will never get a true sense of what drives the company. If, on the other hand, top managers would meet on a regular basis with journalists, financial analysts, and employees, everyone would benefit. Each group would come away with a better understanding of the other's positions. There would be less likelihood of misunderstanding and distrust and surprise. It's a job that no public-relations agency can do without top management's help.

Putting It All Together

Several years ago, a well-known industrialist told me that all business success is based on two things: building relationships and patience.

Nowhere is this more true than in market positioning. None of the market-positioning activities—using word of mouth, developing the infrastructure, forming strategic relationships, selling to the right customers, dealing with the press—will guarantee success by itself. And none of them will bring success overnight. It takes a long time to establish contacts and build relationships.

But taken together, and given enough time, these elements are almost certain to work. They will bring recognition and credibility to a company and its products. It might take a while, but it's worth the wait.

W hen you're rich, they think you really know.

Tevye, *Fiddler on the Roof*

5 Corporate Positioning: There's Only One Thing that Counts

Would you buy a computer from Osborne Computer Corporation? Probably not. The company is still in business, but it went through a highly publicized bankruptcy. Now everyone has doubts about the company—with some justification. Even if Osborne introduced the best personal computer in the world, and developed the best product-positioning strategy, most people would shy away from it. In short, Osborne has lost its corporate positioning.

A strong corporate position is hard to achieve and even harder to regain. Just as product positioning gives individual products a unique presence in the market, corporate positioning provides a unique presence for an entire company.

Corporate positioning is based upon many factors, including management strengths, corporate history, and even the personalities of the top executives. A well-known entrepreneur can help position a startup company. People such as Intel's Bob Noyce and Advanced Micro Devices' Jerry Sanders give their companies a unique personality. Noyce, co-inventor of the integrated circuit, was highly respected in engineering circles, so it was much easier for Intel to position itself as a technology leader.

Similarly, marketing whiz Sanders helped win AMD a reputation as a strong marketing company.

But the most important factor, by far, in corporate positioning is financial success. Without financial success, everything else is meaningless. A company without profits will not maintain its position for very long. When people buy complex products such as computers or telecommunications equipment, they are making a long-term commitment. They don't want to buy from a company in financial trouble.

The flip side of the argument is also true. People feel more secure when they buy complex products from a company with a strong balance sheet. Everyone is eager to listen to profitable companies. As Tevye said in *Fiddler on the Roof*: "When you're rich, they think you really know."

Corporate positioning sits at the top of the positioning hierarchy. Companies must position their products first. Next, the products must gain market recognition. Only then can companies build a solid corporate position.

As the last of the positioning trio, corporate positioning reinforces each of the previous two. A strong corporate position can validate a company's market positioning and its product positioning. When a company establishes a strong corporate position, its other positions become stronger and more lasting.

The Japanese effort in RAM (random-access memory) chips illustrates the point. First, Japanese semiconductor companies made the decision to position their products as high-quality parts. They set up manufacturing operations that minimized the chance for defects in the chips. They also selected a few key customers, such as Hewlett-Packard and IBM, and did additional testing to make sure the chips going to those customers were top-notch.

The Japanese gained market position when Hewlett-Packard began running quality-comparison tests between American and Japanese chips. The Japanese chips had fewer defects in test after test, and HP announced the results publicly. In effect, HP held up its scorecard and showed the Japanese as winners.

After the HP tests, Japanese RAM chips gained market share quickly. Some customers checked other types of Japanese chips and found that they, too, were high quality. Soon, Japanese semiconductor companies had a solid corporate position as high-quality suppliers.

Sometimes, a corporate position can be established on the basis of just one or two key products, which I call "silver bullets." If a company chooses its silver bullets carefully, and highlights them to the public, it can gain a strong corporate reputation, even if the rest of its product line is only mediocre. In this case, image becomes the reality.

Silver bullets are particularly important in technology-based businesses. Products at the forefront of technology are difficult to mass produce. Typically, they sell at high prices and in low volumes. Companies make most of their money on mass-production items, but these products are usually "plain vanilla" products. That is, they are rarely at the cutting edge of technology. Companies sacrifice some performance when they design a product for mass production. Moreover, it usually takes a while to work out all the production problems. By the time the mass-produced product finally reaches the market, new products, using even more advanced technology, already will be out.

The solution, then, is for companies to develop a balanced product mix. They should sell a few silver bullets to build an image of technological leadership, but they also should sell plain vanilla, mass-produced products to keep the money rolling in. Take Intel as an example. It will probably be a long time before Intel's sophisticated 32-bit microprocessors bring in as much money as its 16-bit microprocessors. But the 32-bit chips are widely praised in technology journals. They give Intel an image of technological leadership. That is critical to Intel's corporate positioning. Intel's image of technological leadership is a comfort factor. It reassures customers they are buying from the best.

If a company manages to establish a strong corporate position, or corporate personality, it can reap many benefits. Corporate positioning tends to have long-lasting effects. Consider the case of the Japanese chip manufacturers described earlier. American semiconductor companies have improved on quality, and they now match the Japanese in quality tests. But many customers still believe Japanese chips are of higher quality.

Among the other benefits of an established corporate position are:

Lower cost of sales. When a company has a strong corporate position, the market accepts the company's new products more readily, simply because they carry the company's name. The company's salesmen don't have to work as hard.

Higher prices. Companies with strong corporate positions can sometimes charge higher prices. For many years, Advanced Micro Devices was able to set its prices above the industry average because the company had a reputation for high-quality production.

Faster market penetration. The media usually pays more attention to companies with strong corporate positions.

Customer loyalty. Customers feel more secure doing business with a recognized leader.

Product drag. Not all of a company's products will be world-beaters. A strong corporate position enables the company to be successful in selling its weaker products as well as its strong ones.

Better recruiting. Leading companies can recruit the best talent, because people want to work where the "action" is.

Employee loyalty. A strong corporate character encourages employees to identify with the company's success. It provides focus and direction through the organization.

Higher price-to-earnings ratio. Investors are attracted to companies with a strong corporate position.

Of course, all these benefits will disappear if the company's profits begin to slip. As mentioned earlier, profitability is the most important factor in corporate positioning. The minute profits decline, the market begins to worry. Everything else is called into question.

Tandem Computer provides an example. Tandem was the first company to sell "nonstop" computers—that is, computers that never break down. Ever since it shipped its first computer in 1976, Tandem has built a reputation as a high-quality company. Its technology was advanced and its market positioning was strong.

But when Tandem's earnings dropped below expectations in early 1984, analysts began looking for reasons. They decided Tandem's market positioning and technology were slipping. They wrote reports about new startups that were putting pressure on Tandem. Tandem's hard-earned positioning began to slip away.

The truth is that Tandem's technology and products were just as strong as ever. The new startups were not a threat. Tandem's problems were more complex, relating to forecasting and pricing. The problems were soon corrected, but the damage had already been done. Financial performance clearly had a major impact on Tandem's marketing efforts.

Trilogy Systems ran into even worse problems with its corporate positioning. The company was founded by Gene Amdahl, one of the geniuses of the computer industry. Amdahl has an impressive track record both as an engineer and as an entrepreneur. Amdahl Corporation, the company he founded in 1970 now has sales of $462 million.

So when Trilogy announced plans to develop a line of powerful computers using a new semiconductor technology called wafer-scale fabrication, it was able to quickly establish itself as a technology leader. Raising money was no problem at all, and the company received lots of favorable publicity. But this positioning was fragile. When the company began to have financial problems in 1984, its credibility disappeared. Everyone began to question whether Amdahl could pull it off again.

People don't worry as much about the financial stability of consumer-goods companies. People wouldn't hesitate to buy a tube of toothpaste or a box of detergent from a bankrupt company. Most consumers don't even know what company makes the toothpaste they use.

Buying an expensive technology-based product is a greater commitment. If the company goes bankrupt, who will provide the service? Who will provide the new parts? Once a computer company, such as Osborne or Magnuson or Victor, gets into financial trouble, no one wants to buy their products. No one knows if the company will be around in six months.

When a company loses its corporate positioning, the only thing it can do is to start the whole positioning process over again. It must go back and reinforce its product positioning, then move up to market positioning. Finally, it can try to recapture its corporate position.

Intel went through this process in the recession of 1981-82. Profits fell at all semiconductor companies, and some companies even lost money. Intel had been a darling of Wall Street for ten years, but people began to wonder about the company. Many analysts questioned whether Intel's profit margins would ever return to their prerecession levels. They pointed to the loss of key people and the company's declining market share in the microprocessor market. They even began to question Intel's management practices. People no longer thought of Intel as an innovator or a technology leader.

Intel responded by focusing on products and product positioning. The company introduced more than 100 new products in 1981, a record for the company. Gradually, Intel shifted the focus back to technology and products. When the recession ended, profits returned and people once again began to view Intel as the industry leader.

This focus on financial success is likely to intensify, if anything. Using computerized data bases, people can look up financial information instantly. Say a factory manager is going to buy a sophisticated piece of equipment. He can hit a button on his computer and get a complete financial profile of the company selling the equipment. If the company looks a bit shaky financially, he is likely to buy from a competitor.

I was once involved in a situation very much like that. A friend telephoned me and asked what I knew about Inmos, the British semiconductor company. My friend was in charge of distributing some research funds, and he was thinking about funding a project at Inmos. I went to the computerized data base and got all the public numbers on Inmos. The numbers showed Inmos had been spending a lot of money and losing a lot of money.

I passed along the information, and my friend decided not to fund the Inmos project. He didn't want to fund a company that was losing money. Inmos's technology might have been first-rate, but its corporate positioning was lousy.

Financial officers should keep this story in mind. Positioning strategy is not just for marketing managers. It is for all managers. Financial results can destroy a company's positioning— or they can solidify the company's position as an industry leader.

I f you don't know where you're going, you might end up somewhere else.

Casey Stengel

6 Developing a Strategy: Three Steps to Success

Knowing Where You're Going

In the minds of many, there is a recipe for success in Silicon Valley. The company should design innovative, user-friendly products, using leading-edge technology. It should offer customers a total solution to their needs. It should target Fortune 500 companies as its primary customers. Mix all these ingredients, and sales should grow to $500 million in five years.

Many entrepreneurs think success is as easy as that. Dozens of entrepreneurs approach me each year with business plans not much different from that described above and illustrated in Figure 9. Unfortunately, introducing a new product, or building a new company, is not that simple. Of course, advanced technology and sound management are important. But unless a company can identify and achieve a unique position for its products and itself in the marketplace, it will never succeed. Developing a strong positioning strategy is the key to marketing success.

How can a company develop such a strategy? That is the subject of this chapter. In the last four chapters, I laid out the basic ideas behind dynamic positioning. Now, it's time to look at how companies can put those ideas to work. How can a company identify the proper niches for its products? How can it gauge the trends and attitudes of the market? How can it convert those trends into a successful strategy?

The Entrepreneurial Dream

Market:	Fortune 500
Potential Size:	$500 million in 5 years
Expected Image:	IBM-like
Products:	User-friendly
Design of Products:	Innovative
Technology:	Leading edge
Client contribution:	Total solution
Industry growth:	Dynamic
Competition:	None perceived (creating new business)
Why now:	Window of opportunity
Marketing:	Aggressive, world-wide
Backing:	Kleiner/Perkins, Hambrecht & Quist, Sequoia, Mayfield, A. Rock, Venrock, Sevin/Rosen
Capitalization:	$10 million
Valuation:	$100 million
Headquarters:	Silicon Valley
Management:	Veteran, experienced
Founded:	January 1983
Founder:	Ex-Intel
VP Engineering:	Ex-Hewlett-Packard
VP Marketing:	Ex-IBM
Total employment:	3
Manufacturing:	Low cost
Strengths:	Depth of management
Weaknesses:	None perceived
Why need PR?:	Going public in 3 months

Figure 9

As I see it, developing a positioning strategy is a three-step process. To start with, a company must have a good understanding of itself—its strengths and weaknesses, its goals and dreams. Top managers should have a coherent vision of the culture and goals of the company. If different managers have widely differing visions, the company will never be able to develop a solid positioning strategy.

Second, the company needs to understand the market environment. That is trickier than it might seem. Most companies gather statistics about customer behavior. Then, they make decisions based on the market data. This quantitative approach is quite satisfying to numbers-happy MBAs. In most cases, however, it obscures reality. Instead, companies should use more qualitative approaches to understanding the environment. Marketing managers must develop an intuitive feel for the market. Rather than gathering numbers, companies should listen to customers' frustrations and desires. Their comments won't fit into a graph. But they will lead to a better understanding of the marketplace.

Finally, the company must use all this information to decide on a positioning strategy. There is no single formula for deciding on a strategy. Just as the world is filled with a tremendous variety of technologies and products, so too is it filled with a variety of positioning strategies. Every company must find its own road to success. Managers must keep an open mind and seek a variety of opinions before settling on a strategy. Then, once the strategy is in place, managers should be willing to adjust the strategy as market conditions change.

In this chapter, I'll discuss each of these three keystones of the positioning process: understanding your own company, understanding the environment, and, finally, deciding on a positioning strategy.

Internal Audits: Know Thyself

It is remarkable how many companies have trouble answering the simple question: "What business are you in?" Recently, one of my colleagues interviewed seven people at Cohesive Network Corporation, a startup company in Silicon Valley. She asked that seemingly simple question, and got seven different answers. One person described the company in terms of product applications. Another described it in terms of the technology used in the products. Yet another talked about the nature of the marketplace.

Cohesive was suffering from a severe identity crisis. The company was developing a good product. But inside the company, it had still not developed a coherent identity. Does this really matter? It certainly does. If company executives do not have a uniform and clear vision of where they are heading, they are likely to run into trouble down the road. As the market environment shifts with time, the company might not recognize that it, too, needs to change. Just as bad, company executives could find themselves at odds with one another over how the company should adjust. Infighting could paralyze the company while the industry speeds ahead without it.

Harvard professor Theodore Levitt presented the classic example in his article "Marketing Myopia." Levitt describes the plight of the railroads. Had the giant railroads defined themselves as "transportation companies," they might have adjusted more effectively to the coming of airplanes. As it was, the railroads saw air transportation as competition, not a new opportunity.

Similar situations abound today. Indeed, a clear but flexible view of your company's mission is more important than ever in these times of rapid change. A few years ago, a Silicon Valley company tried to market computer-music software programs. The company went bankrupt when the market did not grow as quickly as expected. The company's problem was that it saw itself as a computer-music company. Had it seen its mission as "creative uses of computers," or even "computer-based entertainment," it would have had a better chance of success. It could have adjusted more effectively to the changing environment. Similarly, publishing companies had best view themselves as "information companies" if they want to survive the next decade.

At the same time, each company must understand the internal "cultural" factors that drive the company. Much has been written about corporate cultures in recent years, and culture can indeed be a powerful force. If a company can develop a culture that emphasizes quality and reliability, all employees are likely to work harder toward those goals. Similarly, a company with a culture that encourages innovation is more likely to develop creative new products.

In new companies, the founding entrepreneurs play a dominant role in determining the company's culture. Long after having given up day-to-day control of the company, Bill Hewlett and Dave Packard continue to have an important effect on Hewlett-Packard through the cultural norms they established at the company. Founders should take time to consider the cultural qualities they want their companies to have. If several founders are involved, they must make sure their ideas of corporate culture are in sync.

Understanding the internal forces at a company should come naturally, but often it doesn't. Many managers are so focused on the outside forces afffecting their companies—new technology, new competitors, new markets—they forget to look inside.

When we begin to work with a new client, the first thing we do is to force the company to do some soul-searching. We have developed a formal process called an "internal audit," during which we probe the inner workings of the company to find out what really makes it tick, and to make sure that different parts of the company aren't working against one another.

Typically, we meet with five or six top people in the company. We always talk to the president and the vice president of marketing, and we usually meet with the vice presidents of engineering, finance, and manufacturing as well. Each interview lasts for ninety minutes or so. During that time, we ask general questions about the company's history, its products, market, competition, and goals. We also ask about the individual himself, and how he perceives his role in the company.

We perform these audits in part to acquaint ourselves with the new client, and to gather information for upcoming communications efforts. But more important, the audit helps in developing a positioning strategy. The perceptions and attitudes of company executives are critical to the company's positioning. At National Semiconductor, for example, chief executive Charlie Sporck is a manufacturing-oriented manager. His emphasis on manufacturing flows down through the company. So it was only natural that the market would come to view National as a production-oriented company, a company that could turn out high volumes of product at low cost.

By examining the perceptions and attitudes of individual managers, we also can expose inconsistencies and conflicts among company executives, and we can get a feel for the company's expectations. After finishing the internal audit, we typically conduct an "external audit" of key industry observers and potential customers to determine whether the company's goals and expectations are, in fact, realistic. (External audits will be discussed in greater depth later in this chapter.)

Obviously, the questions we ask in an internal audit will vary from one company to another, and, to a lesser extent, from one individual to another within the same company. You can get some idea of the process from the following questions, used during the internal audit at a software company:

- What business are you in?

- Describe your market. Is it mature?

- What do you think are your company's strengths and weaknesses? Technical, financial, cultural, others?

- What do you think the public considers to be your strengths and weaknesses?

- Who are your competitors? What are their strengths and weaknesses?

- What directions do you foresee taking in the near future and longer term? That is, what is your market strategy regarding operating systems, languages, and applications software?

- What are the key factors for success in each of the above market segments?

- How long will it take you to implement your strategy in each segment?

- What are the trends in each of these market segments?

- What percentage of the company's resources will be devoted to each market segment?

With these and other questions, we are not trying to gather facts and figures about the company. Instead, we are looking for qualitative information, such as perceptions and feelings. We try to avoid making any assumptions ahead of time, and we lead clients only to stimulate their own thinking. Oftentimes, we will let clients "ramble" in order to learn more about their hidden feelings. Most important, we want to find out if all the key members of the management team are singing the same song, and how deeply they believe in that song.

Even with simple questions, an internal audit can expose important information and insights about the company. Take, for example, the final question in the above list. In estimating the percentage of resources that should go to the development of language software, executives at the software company gave consistent answers. All the responses fell between 20 percent and 30 percent. But for operating systems, responses varied from 25 percent to 75 percent. Estimates for consumer-market software also varied widely.

These differing estimates exposed an important split in the company. A more qualitative analysis of the executives comments backed up this finding. The company seemed to be split into two camps that, in many ways, were sabotaging each other. One camp, led by a manager who favored consumer marketing in the style of Procter & Gamble, wanted the company to focus on entertainment software. The other, led by a manager with a strong technical background, wanted the company to dedicate its efforts to systems software—a smaller market, but one in which the company had already proved itself.

An internal audit at another company, this one specializing in communications, unearthed a different type of conflict. We found the engineer in charge of software development thought he was the top engineering officer at the company. That would have been fine, except the engineer in charge of hardware development thought the same thing.

Most companies, especially small and fast-growing ones, like to sweep these conflicts under the rug. They ignore conflicts until company performance starts to sag. That is a formula for trouble. It is critical each member of the management team understands his or her own role in the company, and that they all share a common vision of the company's plans and goals.

To develop that common vision, different parts of the company must understand each other's needs and goals. In particular, marketing and product-development staffs need to work closely together and grow to understand one another. At many technology-based companies, these staffs hardly interact at all. Marketing people develop positioning strategies without even consulting the product designers, and certainly without any deep understanding of the product-development process.

This compartmentalization of companies, the splitting of marketing and product development, is a serious problem. As advanced technologies influence more and more industries, marketing managers everywhere need to learn to package and symbolize the essence of a technology. They must take the complexity of a technology and turn it into simplicity. They must remove the customer's fear and create an aura of familiarity around the technology.

This is a difficult job, with objectives that sometimes seem incompatible. Marketing managers must let the customer know the product is technologically advanced and exciting. They must not denude the product of its technology. People like to have the latest technology. The product should be a symbol of advancement, a type of status symbol.

Intel's Ed Gelbach once called this the "hexachlorophene syndrome." In the 1950s, a toothpaste maker advertised a secret ingredient called hexachlorophene, even though few consumers understood the term. Today, customers are excited to own a "16-bit processor," even if they don't know the difference between a bit and a byte. At the same time, customers want to be reassured about the new technology. Marketing managers must find ways to make the customers feel comfortable and secure with new ideas and new technologies.

To translate complex new technologies into the appropriate symbols and messages, marketing managers must understand and appreciate the technological process. Without that type of understanding, it is impossible to create marketing messages that capture the importance and meaning of the product. I still find it useful to go to a new production line and watch the products—chips or computers or video screens—coming off the line. It helps bring the technology to life.

Even better, I like to see the product at early stages in the design process. I remember seeing Intel's 8086 chip at a very early stage. The drawing of the layout filled an entire wall. At

Genentech, I saw a photograph of DNA strands and was told these strands were the beginning of a chain that would produce pure human insulin. In each case, I gained a sense of the complexity of the product, and I was better able to communicate the excitement of the new technology.

Designers themselves can offer important marketing insights. In fact, I would say I have gotten more good ideas from technical people than from marketing people. Technical people have a lot of intuitive knowledge that doesn't come across on the spec sheet. Of course, some engineers are not very articulate. But others can speak about technology in an almost poetic way. They have lived with the product and understand the technology on a deep level. They can explain why the technology is important, and where it is heading.

Companies where marketing people don't mix with designers often miss out on great opportunities. The marketing people don't really understand what their company's products and technologies are all about. Xerox fell into this trap and the company has suffered as a result. Researchers at the company's Palo Alto Research Center have been at the forefront of the personal-computing revolution. They developed many of the technologies that Apple used in the Macintosh computer. But Xerox marketing managers never had an appreciation for these advanced technologies. Xerox introduced many of the technologies—such as the mouse and bit-mapped display—in its Star computer two years before Macintosh, but Xerox's marketing managers did not position the product well. They didn't realize what a powerful technology they had. Several frustrated Xerox designers defected to Apple, helping Apple turn those same technologies into a mass-market success.

External Audits: A Qualitative Approach

No company can develop a positioning strategy in a vacuum. No matter how advanced its technology, no matter how low its prices, no matter how extensive its distribution, a company will succeed only if it understands the market environment into which it is selling its product. That is, it must understand the strengths and weaknesses of its competitors, the perceptions and attitudes of potential customers, and the social and political trends of the nation. A product that flopped miserably might have been a wild success had it been introduced six months sooner—or six months later. It all depends on the market environment.

Companies must satisfy customer needs, not simply produce goods. And to do that, they must monitor and understand the environment. For companies in fast-changing industries, this task is particularly difficult and important, as the environment is on a roller coaster of change. Only with constant and creative monitoring of the environment can these companies position their products effectively.

Most companies survey the market to determine the level of customer demand. But the qualitative aspects of the market are more important. By monitoring the market, a company can learn how receptive customers are to change, and what "mental obstacles" the company must overcome to get people to accept new products and technologies. It can get a better sense of customers' expectations, their level of understanding, and their willingness to be educated.

A few years ago, manufacturers of digital control equipment found that workers in mature process industries, like the tobacco industry, felt uneasy about new digital equipment. Why? Interviews revealed the workers were accustomed to operating equipment that had knobs and gears. To operate the new digital equipment, workers had to push buttons, and workers felt uncomfortable pushing buttons. They would rather turn knobs as they always had. So manufacturers began installing knobs rather than buttons on their equipment.

When Apple introduced its first personal computer in 1976, it also had to account for customer perceptions and attitudes. Apple's founders were keenly aware of the public's fears of computers, and they took steps to ease those fears. They chose the name "Apple" because it was friendlier than "Computec" or "Alphatecdyne." By developing a colorful logo, they made the machine seem even less intimidating.

Management expert Peter Drucker has noted that companies cannot really answer the question "What business am I in?" until they understand customer attitudes and perceptions. He explains:

> *W hat is our business is not determined by the producer, but by the consumer. It is not defined by the company's name, statutes, or articles of incorporation, but by the want the customer satisfies when he buys a product or service. The question can therefore be answered only by looking at the business from the outside, from the point of view of the customer and the market.*

In an attempt to understand customer wants and needs, most companies use market research and statistical analyses. In some mature businesses these techniques work adequately. Tire manufacturers, for example, have developed statistical demand analyses that link tire sales to automobile sales. And manufacturers of toothpaste can forecast demand largely on demographic data showing the number of people in each age category.

However, these statistical techniques do not work well in fast-changing industries that are moving into unexplored territory. When you are creating new markets, no one really knows where you are headed. You have to be more creative. John Sculley, the chief executive at Apple, says he is wary of numbers-oriented analysis. He explains: "The only quantitative data I use are what people have done, not what they are going to do. No great marketing decisions have ever been made on quantitative data."

Indeed, market statistics are rarely meaningful in rapid-change industries. Look, for example, at the personal-computer software industry. Projecting the size of this industry is little more than guesswork. In 1983, three respected market-research organizations tried to project growth for the personal-computer software industry. The three companies couldn't even agree on the size of the market when they looked backward, to 1982. When they tried to project the industry size in 1987, the numbers flew off in all directions, ranging from $3.7 billion to $13.6 billion. What can you conclude from these numbers? Not very much. You can choose any numbers you want, and reach any conclusions you want.

This case is hardly unique. In growth markets, few projections have ever been correct. Even the Semiconductor Industry Association, which gathers data directly from the companies in the semiconductor industry, is usually way off target with its projections. Surveying customers doesn't help much. Customers are always enthusiastic about future products, but they don't necessarily buy the products once they are introduced. In many cases, there seems to be an infinite demand for the unavailable.

Nevertheless, market projections proliferate. Like economists, market researchers do not seem shamed by their continuing failures. Numbers seem to make people feel more secure, so everyone tries to come up with some statistics, however meaningless they may be. Rare is the marketing vice president who has the confidence to go into a meeting and say: "There simply aren't any good numbers available." Instead, marketing managers somehow create numbers that justify their plans.

Numbers with little scientific basis swirl around the industry and sometimes come full circle. When *Business Week* did an article on commercial uses of superminicomputers in 1979, there were no projections of the market. So a *Business Week* reporter called three companies and produced a graph for the magazine in thirty minutes. A few weeks later, the reporter was working on an article about computer printers. A leading manufacturer of printers showed him the *Business Week* graph and used it to justify his company's strategy and production plans.

The problem is caused, in part, by the way business schools train MBA students. Students are taught to rely on abstract theories and numbers-oriented planning. They have little appreciation for how fast-changing industries really work. Tektronix, a leading manufacturer of electronic instrumentation, learned this lesson the hard way a few years ago. Recognizing that it needed to become more marketing-oriented, Tektronix hired a group of new MBAs with interests in marketing. They came to Tektronix armed with charts and theories, but they only made matters worse. They had no understanding of the peculiarities of the electronic-instrumentation business, and they never made much of an effort to learn. Within a few years, nearly all the hot-shot recruits were gone. Tektronix still needed marketing help, but it learned that a flock of MBAs was not the answer.

The case-study approach used at many business schools causes additional problems. In the case-study approach, students learn by analogy. But for many new markets, there are no analogies. Personal computers don't follow the same rules as stereos or consumer electronics. They don't even follow the same rules as big computers. The textbook has to be rewritten every day in emerging industries.

The problem with most traditional approaches to marketing is that they try to accurately predict the unpredictable. Unfortunately, you can only measure what you can control. And you can't control how people will respond to new developments. People themselves don't know how they will respond to new developments. Before 1984, many polls tried to predict how people would react to a woman vice-presidential candidate. But these polls were meaningless. Until people were actually confronted with Geraldine Ferraro, they could not answer in a meaningful way.

Trying to make predictions about technology-based products is even tougher. In 1976, very few people thought they needed a personal computer. Traditional market research would have shown a market of a few hundred, not a few million, units. Only as companies brought new and useful software to the market did people realize a personal computer could, in fact, be a useful tool.

Recognizing this problem, some leading-edge companies pay no attention to market research. At Activision, a leading producer of video-game software, no one does any market research until the game is finished. James Levy, Activision's chief executive, explains it this way: "Market research will kill as many good games as bad ones. It's not a definitive tool. You never know about a new title for sure until it hits the street."

Levy believes companies in fast-changing industries must wean themselves from numbers. "People want everything to be predictable," he explains. "They try to turn everything into a science. They're uncomfortable with uncertainty. But to be successful in this business you must deal with and live with uncertainty and surprises. A certain amount of uncertainty you must accept as a fact of life."

Indeed, it is insecurity that drives the development of new markets. Insecurity keeps companies sensitive to changes. Statistical information, with its illusion of certainty, provides a false sense of security. Companies relying on statistical data become lax and less sensitive to changes in the market. In fast-changing markets, you can't afford that.

If marketing managers can't rely on numbers, how can they ever understand the market environment? They must use a combination of intuition and keen sense for changing attitudes. They must develop a feel for the market, just as a card player develops a feel for the table. That doesn't sound as scientific as statistical analysis, but the fact is, it works better.

Developing a feel for the market, an intuitive sense of the market, is not easy. Few people can articulate how they do it. A good salesman can't necessarily explain his technique, but he'll know exactly when to close an order. Similarly, an effective marketing manager will say: I just sense this is the way the market is.

You get that type of sixth-sense only by spending time in the marketplace. You need to live and breathe the market. You need to talk to market participants on a continuing basis. It is ironic, but true, that in this age of electronic communications, personal interaction is becoming more important than ever.

Some product marketing people spend their time putting together data sheets, writing application notes, conducting training sessions, sending memos to the field. But memos do not a market make. In my mind, marketing people should be on the road half the time—meeting customers, talking to people, building relationships, seeing where the next product should go.

Indeed, conversations with market participants often provide more insights than a long list of statistics or a set of sophisticated theories. In marketing, experience is more valuable than logic. Students coming out of Harvard Business School think they're going to teach the world how to market. But experience is far more valuable.

In most aspects of Western life, logic is king. The Western approach to life encourages us to break things apart and analyze them. It embraces an engineer's view of the world: Things that can't be measured are irrelevent or illogical. This approach underrates the value of intuition. When you break things into little pieces, when you "statisticalize" them, you lose the intuitive feel that is so important. When you turn a perception into a statistic, you rob the perception of its richness.

Relying on intuition scares people. But the most successful marketing people rely on intuition. John Sculley of Apple says most of the important marketing decisions in his life have been intuitive decisions. Intuition should not be seen as negative. It is merely another form of knowledge. Intuition comes largely from experience, rather than intellectual or analytic thinking. You gather information through your senses, sort through it in your subconscious mind, and intuition emerges. It might sound "soft," but it is just as valid as any other form of thinking.

In an intuitive approach, it is important to look for patterns and trends and connections, not raw numbers. Statistical studies might show 10 percent of your customers are disappointed with their products. That doesn't tell you whether the 10 percent is growing or shrinking. Nor does it tell you how intensely the disappointed customers feel. Or whether the disappointed customers are influential and vocal.

Focus groups can provide that type of qualitative information. In these groups, potential customers directly express their ideas and opinions. At too many companies, however, the information from focus groups is quantified before it is distributed to decision-makers. This filtered information is not nearly as useful in understanding the attitudes and perceptions of customers. Key decision-makers must hear the opinions of customers directly.

Qualitative information can come from all types of sources. I have gained lots of useful information just by standing at the counter of an electronics or computer store. In the early days of calculators, I saw a man take two calculators in his hands to see which was heavier. He bought the heavier model. So I suggested to a client company that it put weights in its calculators. It did, and that seemed to help sales. The weights made the calculators feel more substantial.

You don't get information like that by going out and asking 10,000 people to list their likes and dislikes. You do it by observing. When you go to enough retail stores, you get to understand the selling process. You see what customers are asking about. You see where they hesitate. You get to understand their fears.

Many marketing managers get so wrapped up in their products they become deaf to criticism and insensitive to the market. They spend all their time promoting the strengths of their product. They begin selling everybody and listening to nobody. They begin to think their view is the view of the marketplace. A marketing manager at Spectra-Physics, who recognized this problem, told me he got rid of one outside marketing consultant because "he began to believe us." It is critical to maintain a fresh and unbiased view of the marketplace. That comes only by talking, and listening, to customers.

When you go out and ask questions, you don't need to talk to a "statistically significant" sample. You just need to talk to the right people. I recently did some work for a fast-growing telecommunications company. Rather than conducting a survey of 500 people, I talked to the telecommunications managers at Coca-Cola, McGraw-Hill, and a dozen other major companies. I quickly discovered the key issues affecting telecommunications customers. Marketing managers should make regular trips to customer sites. It is probably one of the best market research tools that exists.

We use these ideas when we perform "external audits" for our clients. The purpose of an external audit is to gather information and insights from the environment. That information is then used in developing a positioning strategy. The audit can act as a "reality check" on the perceptions that were expressed by company executives during the internal audit described earlier in this chapter. Sometimes, the external audit shows the executives understand the market well. Other times, it shows company executives are out of touch with the realities of the market, so positioning plans must be drastically altered.

During the audit, we interview people from a number of different groups: existing customers, potential customers, distributors, industry "experts," financial analysts, and perhaps key journalists. We typically talk to a dozen or so people in all. For a semiconductor company, we might talk to five startups that will be buying semiconductors, five established semiconductor customers, and five industry analysts.

In these interviews, we do not look for specific facts and figures. Usually, we do not even talk about the specific product.

Rather, we identify patterns, attitudes, and opinions that influence the thinking process. For example, we want to find out how open people are to the acceptance of new technology. Are they willing to try something different and new?

We might ask questions such as:

- Of the products currently on the market, which do you like best? Why?

- Where do you think the market is headed? What are the most important trends?

- What do you think of XYZ technology? What are its advantages and disadvantages when compared to ABC technology?

- Which companies do you see as the rising stars of the industry? Why?

- When you buy this type of product, what factors influence your decision? How much are you influenced by cost? Ease of use? Added features?

- What do you see as the major limitations to growth in this market?

- Who do you see as the key opinion leaders in this industry?

A recent audit we performed for Rolm provides an example. Rolm was about to introduce an advanced telephone exchange system that would handle both voice and data. We talked to fifteen or twenty managers in charge of either management information systems or telecommunications for their companies. Among the questions we asked: How soon do you see the convergence of voice and data? Who do you see providing that kind of solution? Who at the corporation will be responsible for purchase decisions—the MIS manager or the telecommunications manager? Will the breakup of AT&T have a positive or negative impact on this trend?

We found there were many divergent views. Some thought integrated voice-data systems were twenty years away. Some said integration had already begun and they needed it yesterday. There was a tremendous amount of confusion in the marketplace. The breakup of AT&T added to the confusion. As a result, we decided Rolm had to simplify the message associated with its new product. The company had a good product, but the product and the explanations of its advantages were quite complex. That complexity would just add to the confusion. With a simplified message, Rolm had a great opportunity: It could stand above all of the confusing noise in the market.

Sometimes, external audits simply confirm what the company expected. An audit for Cohesive Network, for example, showed that telecommunications managers were increasingly interested in using private networks. In other cases, the audits show the company's perceptions are somewhat off-base. Convex, a startup in the supercomputer business, thought the key feature that distinguished supercomputers from smaller machines was their ability to handle sixty-four bits of data at once. Instead, customers believed the ability to perform integrated vector processing was the key feature. Convex also discovered a new market to target. It expected people involved in seismic exploration would be most excited by low-cost supercomputers, but it found there was just as much excitement among people in the computer-aided design business.

External audits are particularly useful when they reveal trends and patterns for which the company was not even looking. To help in positioning a new microprocessor for Intel, we talked to engineering managers at Hewlett-Packard, Xerox, TRW, and several other major companies. We asked about their expectations for the next generation of microprocessors. We found significant segmentation at age thirty-five, even though we weren't specifically looking for it. The managers older than thirty-five were reluctant to try a new technology. They were more interested in quality assurance and documentation. The younger managers, on the other hand, wanted to experiment. We learned that Intel should present different messages to different managers, depending on their ages.

External audits should not be a one-shot deal. Companies must continuously monitor the environment to detect changes in mood and attitude. Everyone, not just marketing managers, should be involved in the process. Engineers should meet regularly with customers, and so should top management. Only through constant monitoring of the environment can companies stay on the right track.

Many startups forget about this as they grow. Entrepreneurs usually start off on the right foot. They almost always have an intuitive feel for the markets to which they are selling. Top executives deal directly with their customers. They are constantly aware of what is happening in the market. This is a key reason for the success of so many small companies in fast-growing industries. They are more in tune with the market than their larger competitors.

As they grow, corporations tend to lose this feel for the market. They begin to suffer from what I call "bigness mentality." Top executives forget their roots. Rather than relying on intuition, as they did in the early days, they start to manage by numbers. Qualitative information becomes statistical. As staffs grow, they separate the top managers from the market. Managers begin to worry more about the efficiencies of mass production and less about the needs of the market. They no longer want to take risks. Their whole thinking process changes.

To continue with success, growing companies must continue to "think small." Managers must maintain their intuitive feel for market trends and attitudes. They should look at the numbers, but they shouldn't be ruled by them.

Deciding on a Strategy

Once a company understands the market environment, it must decide on a strategy to get its products positioned within that environment. This is a fuzzy process with no firm rules. It is different in every case. Sometimes, it evolves naturally. Other times, it is the result of formal meetings and strategy sessions. In the best cases, it is a combination of the two.

Coming up with a massive positioning document is not all that important. Marketing plans usually sit on shelves, gathering dust. People don't follow them day to day. Rather, it is the ideas that are important. The key people in the company must converge on a common positioning strategy, then put it into effect. Spending weeks or months writing down the ideas is simply a waste of time.

To give a taste of the process, I will describe how we conduct positioning sessions with our clients. The sessions have two primary goals. First, to identify a position. And second, to decide what actions are needed to achieve the position. As a result of the sessions, many things can change. It is not just a matter of coming up with a new slogan. The company might change its market direction, its target customers, its distribution strategy. It might even change the products themselves.

We run positioning sessions after completing the internal and external audits, so we already have a good feel for the internal dynamics of the company and the attitudes of the marketplace. We walk into the meeting armed with direct, qualitative information from the marketplace. The challenge is to use this information to help the company target its product and develop a unique position in the marketplace. We must decide how to differentiate the product from its competitors, how to distribute and promote it, and how to gain credibility in the marketplace.

Typically, we meet with six to ten people from the company. The people represent different groups and different experiences. Some people should be from the sales organization. They continually interact with the customers, so they have direct information from the market. But their views are somewhat limited: Selling is a one-to-one process, and marketing must consider the broader picture. The meeting also should include some people from the technical side of the company. They, too, have a somewhat narrow perspective, but they can mark the boundaries of the discussion. They won't let you go too far with your analogies.

Each person can relate individual experiences. One might say: "When I was at DEC, such and such happened." Another will add: "When I was at Hewlett-Packard, we did it this way." With a new product, none of these past models will be perfect. But some mixture of these experiences should lead to the right approach.

The interaction at the sessions usually is not systematic. It's more like brainstorming, or free association. The goal is to maximize creativity. Everyone tosses out ideas, then others modify the ideas and add to them. In running a session, I try to listen to a variety of experiences and a diversity of views, without allowing any one of them to dominate my thinking. I make sure not to reach conclusions too quickly. I don't want to get locked in place and lose flexibility. Once you draw a conclusion, you begin to argue over the conclusion, rather than look for new insights. The phenomenologist philosopher Husserl used to say: Bracket your prejudices. I try to follow that advice. I keep prejudices out so reality can penetrate.

Ideally, the session breaks neatly into three stages: Input, Analysis, and Synthesis. Each stage lasts about an hour. In the first hour, I sit and listen. I take notes and absorb information. I look for patterns and connections. These don't always come quickly. I wait, wait, wait. Maybe there's a different way. Maybe there's a better way. I encourage off-the-wall ideas. I look for things that the other people know intuitively but they haven't been able to articulate.

I'll look for all types of relationships. How does the product relate to past and future products? How will salespeople and customers relate? How does the operating system relate to the applications program? How does the telecommunications manager relate to the MIS manager? How does the company relate to its suppliers? I'm always looking for ways in which the company can use these relationships to its advantage.

In the second hour, I write some of my ideas on the board. I list obstacles. Competitors. Environmental factors. I put all types of things on the board and try to relate one thing to another. I'll bring up examples from other companies, add some ideas about industry structure.

Sometimes, I'll use the idea of "positioning denominators." I make a chart listing strengths and weaknesses in the three positioning categories: product, market, and corporate. In the product category, I compare the product to its competitors in terms of power, speed, compatibility. In the market category, I compare distribution, sales forces, customers. In the corporate category, I compare financial resources, reputation, management image.

The third hour of the session is devoted to synthesis. There are no numbers or graphs, just manipulation of ideas. Working together, we try to integrate all the ideas brought up by the salesmen, technical people, and others. We link together the strengths among the positioning denominators, and try to turn them into a coherent plan.

We battle back and forth until we hit upon the right positioning plan. It usually comes in an "aha!" experience. All of the
sudden, everything makes sense. Everyone agrees: "That's it!
That's right!" Out of the murky mess of information, a clear
vision of the future has emerged. I've gone through the process
hundreds of times, and we've reached that type of conclusion
in all but a couple cases.

One of the failures involved Imagic, a company that produces
video-game software. We met with Imagic managers for a total
of six hours, but we could not find any perceptible differences
between their strategy and the strategy at industry leader Activision. None. They simply said: "We're going to make games
and take market share away from Activision." Their goal was to
get a lot of sales and to go public fast. There were no real
differences between Activision and Imagic. If one had a "Turtle
Walk" game, the other would have a "Rabbit Walk" game. The
differences were all superficial. Imagic had no solid basis for
differentiation of themselves or their products.

In most cases, though, the combination of internal audit,
external audit, and positioning session leads to a clear positioning statement. Convex, the supercomputer company mentioned
earlier, provides a good example. The company has a strong
technology group, led by Steve Wallach, one of the star engineers
in the book *Soul of a New Machine*. When they came to us,
however, the company founders had no clear vision of their
position in the market.

Convex initially planned to position its computer as a superminicomputer. That product category was established in the late
1970s, when Digital Equipment Corporation introduced a line
of computers, known as VAX computers, that was more powerful than traditional minicomputers. Within a few years, superminis became very popular, particularly for scientific
applications. Convex was designing a machine that would run
the same software as a VAX, but would be even more powerful.

There was one big problem: The superminicomputer field was
already quite crowded. There were more than fifty companies
selling superminicomputers. Convex's machine was probably
better than the rest, but it would be difficult to make the machine stand out in that market.

So we began our external audit and looked for ways to differentiate the Convex computer. We found a tremendous demand for superminis like the VAX, but also growing dissatisfaction. After many years as the workhorse for scientific applications, the VAX was starting to show its age. Scientific problems, such as the design of very-large-scale integrated circuits, were becoming more complex, and VAXs could no longer handle the job very well. Many users wanted the high speed and special capabilities of a "supercomputer." But existing supercomputers, sold by Cray and a few other companies, had drawbacks. They were too expensive for most applications. And they couldn't run nearly as many different types of software programs as the VAX could.

At the positioning session, we quickly recognized there was a huge gap in the market. On one side was the VAX: lots of software, prices ranging from $500,000 to $2 million, but insufficient power for many new scientific application. More than 50,000 of them had been sold. On the other side were Cray-like supercomputers: lots of power, but not much software and prices up to $5 million to $10 million. Fewer than 150 of these computers had been sold.

We saw an opportunity to open an entirely new market segment between the VAX and the supercomputers. There was a huge gap in price and performance, and the Convex computer could fit right in the middle. It was twenty times faster than a VAX, but only one-quarter the price of a supercomputer. It could run all the VAX software, but it also could provide almost as much power as a supercomputer. In the future, the computer's performance could be incrementally expanded into a more powerful supercomputer.

The question then became: Should we position the computer as a super-VAX or baby supercomputer? We quickly decided that it made more sense to position the computer as a baby supercomputer. Rather than competing directly against fifty suppliers of VAX-like machines, including powerful DEC, Convex would be positioned in the supercomputer industry, a segment with only three or four manufacturers.

The technology had not changed at all, but the marketing plan was now totally different. The company has begun to think of itself differently. It is in a Cray-like business, not a DEC-like business. That means different types of pricing and different types of marketing. Convex managers now pay attention to all issues relating to supercomputers. For example, the press has become quite interested in the Japanese efforts to leapfrog past the United States in supercomputer technology. Now Convex can share in that limelight. Convex's president was invited to participate in a round-table discussion on the Japanese challenge, giving the company unexpected visibility. Suddenly, Convex is not just an innovative company; it is an American asset in an important international competition. It's amazing what a little positioning strategy will do.

Startups are not the only companies that need to formulate positioning strategies. In dynamic markets, companies must constantly re-evaluate their positioning plans. A positioning plan that seems to make sense one month can be thrown into disarray the following month as new products come to the market and customer attitudes change.

Intel faced this problem with its highly successful 8086 microprocessor. After its introduction in 1978, the 8086 quickly became an industry star. It gained a dominant market share in the market for 16-bit microprocessors—that is, microprocessors that can handle sixteen bits of data at one time.

By 1981, however, storm clouds were gathering. Motorola had introduced a competitive chip in late 1979, and the Motorola chip, the 68000, was beginning to attract attention in the industry. Motorola had followed a classic "second-company-in" strategy. It improved on some of the weaknesses of Intel's 8086, then tried to grab onto the market momentum the 8086 had created.

Intel's marketing staff had been slow in identifying the new trend. They were sitting in Santa Clara looking at industry growth charts and design wins, and they didn't see any big problems. But Intel salesmen in the field began to see something different. The 8086 was still selling well, but customers were clearly intrigued with the Motorola chip. The salesmen had to work harder to sell the Intel 8086. They sensed momentum was shifting from the 8086 to the 68000.

Clearly, Intel needed to do some repositioning of the 8086. This was more than a chip v. chip battle. Entire product lines were at stake. If a microprocessor sells well, it emerges as a standard and all the other chips in the product family are carried on its commercial coattails. The battle between the 8086 and the 68000 was really a battle between Intel and Motorola for industry leadership.

To get a better feel for customer attitudes, I did a quick external audit of the market. Then Intel president Andy Grove called together a special committee to study the problem. The committee consisted of six Intel managers and myself. We met for three days straight, from Wednesday through Friday, at Rickey's Hyatt House in Palo Alto.

Our repositioning project went by the code name CRUSH. Our mission was straightforward: Identify why the Intel 8086 was beginning to slip in its competition with the Motorola 68000, then implement a strategy to respond and recover.

We began by dividing the market into different types of customers. We segmented the customers not according to size or location, but according to their thinking process and attitudes. We decided that some customers were hardware oriented. They cared most about raw performance factors, like speed and power. Other customers were software oriented. They had much different priorities. They wanted a microprocessor with a "clean" architecture so it would be easy to develop software for it.

The external audit, along with comments from Intel salesmen, indicated the Intel 8086 was holding its own among hardware-oriented companies. But the Motorola 68000 was gaining quickly among software-oriented companies. Software developers felt more comfortable with the Motorola chip. They felt it provided more support and flexibility for developing new applications. We didn't have specific statistics to support these findings, but our qualitative information left little question in our minds.

The challenge, then, was for Intel to reposition itself among software-oriented companies. We decided on several ways to do that. One way was to focus more on the breadth and depth of Intel's product line. As long as customers continued to focus on chip v. chip, 8086 v. 68000, Intel would have troubles. In a horse race among chips, Motorola would win among certain customers. But if customers looked at overall solutions and future directions, Intel would have advantages. Intel's 8086 could be combined with its 8087, for example, to provide the best solution for scientific applications. It could be combined with a different Intel chip for a communications application.

By focusing on the whole product line, we also hoped to get customers thinking about the future. We wanted people to worry about the consequences of committing themselves to Motorola. We wanted to play on the customers' fears. Sure, Motorola had a hit product, but could the company support it with other chips and future enhancements? The 68000 had almost no software, no peripheral chips, no development system. And Motorola had not explained its future plans. By committing to the 68000 architecture, might customers get stuck in the future?

Intel, by contrast, already had a full family of microprocessor products. It was a safe bet for the future. To reinforce this point, we planned to show customers Intel's plans for future-generation microprocessors, both up and down the product line. The message would be clear: Intel had a well-developed plan for the future. With Motorola, the future was murky.

We also identified another Intel advantage. Intel's top executives—Bob Noyce, Gordon Moore, and Andy Grove—were perceived, accurately, as pioneers and innovators in semiconductor technology. Their credibility was high. If they talked directly to major customers, their message would carry a great deal of weight. We planned to have the top three and other leading technical people make presentations at small seminars. These seminars would require a lot of valuable time, but they would make a much stronger impression on customers than advertisements and articles. Intel would come across as having a great deal of technological depth, just on the basis of the people who gave the seminars.

Intel wasted no time putting CRUSH into action. Our group finished its three-day positioning session on Friday. The following Tuesday, the group presented its findings and requests for budgets to the executive staff of Intel. On Wednesday, we assembled more than 100 Intel managers from all over the world to explain the project. Each was assigned a specific task—a software task, a technical task, a documentation task, an advertising task.

It took Intel less than seven days to develop a new positioning strategy and put it into place. This ability to respond quickly is an important corporate asset. Sometime later, when I told a former Motorola executive that it took only seven days to develop the CRUSH program, he told me that Motorola could not have even organized a meeting in seven days.

During the next three months, Intel executives gave presentations to more than thirty major customers. In the following quarter, Intel gave nearly fifty half-day seminars to potential customers. Gradually, the momentum shifted away from the 68000 and back toward the 8086. Motorola had a strong technical product, and its sales continued to grow. But Intel had won the positioning battle. Its 8086 remained the leading 16-bit microprocessor in the industry.

There is no special magic in what Intel did. This same approach can work for other companies and other industries. Many companies lose their "capacity to act," losing both the value of the market momentum and the timeliness of response. Companies that plan qualitatively and react swiftly always will be a step ahead of the competition in the battle for strong market positions.

W*e have met the enemy and he is us.*

Pogo

7 Why Marketing Plans Fail: The Ten Competitors

Intangible Competition

Ask a marketing manager to name his primary competitors, and he'll rattle off the names of a few other companies in the industry. Marketing managers in the personal-computer industry worry about competition from IBM and Apple. Those in the semiconductor industry worry about Intel and Motorola.

These worries are, to a large extent, misplaced. Certainly, IBM, Apple, Intel, and Motorola are all tough competitors. But they aren't the toughest competition. They aren't the *real* competition.

The real competition comes from what I call "intangible competitors." These competitors involve ways of thinking and ways of looking at the world. When a marketing manager resists change, that is an intangible competitor. When an entrepreneur begins thinking in the bureaucratic style of a large-corporation man, that is an intangible competitor.

These intangible competitors are the primary reason marketing plans fail. If companies can deal with these competitors, they are bound to succeed—no matter what other companies in the industry do. I have identified ten intangible competitors that all companies confront, regardless of what industries they are in. They are:

1. **Change**

2. **Resistance to change**

3. **Public knowledge about the product**

4. **The customer's mind**

5. **The commodity mentality**

6. **The bigness mentality**

7. **Broken chains**

8. **The product concept**

9. **Things that go bump in the night**

10. **Yourself**

Competitor 1: Change

Our society is in a perpetual state of change. Everything is changing.

Companies change. One day the newspapers carry a story about a computer company hitting $100 million in sales. A few weeks later, they carry a story about the same company going bankrupt.

Industries change. The breakup of AT&T is radically tranforming the communications industry. A few years ago, AT&T had a monopoly on long-distance service. Now there are dozens of competitors. The software industry has undergone an even bigger change. A decade ago, the industry included a few hundred companies. Today there are thousands.

Products change. Today, it seems that every product is becoming "smart." Microwave ovens have microprocessors in them. Telephones have microprocessors in them. Even toys have microprocessors in them. With these microprocessors tucked inside, familiar products take on new traits and perform new tasks. Computers themselves are changing too. Today we have computers in all shapes and sizes—personal computers, hand-held computers, portable computers.

Distribution channels change. A decade ago, no one believed you could sell a computer through a retail store. Today, retail stores sell hundreds of thousands of computers each month.

Issues change. A few years ago, every newspaper and magazine was writing articles about the high quality of Japanese products. Now, we hardly hear anything about that. The media is in a state of perpetual motion, latching onto a "hot" topic one day, then forgetting it when a new issue appears on the scene.

These changes are a major competitive force. They have a deep influence on the growth and direction of every company. Companies blind to change are doomed to failure. Change can topple even dominant companies. Companies simply cannot afford to stay in the same place.

Business history is full of examples of companies that didn't recognize change in the market, and paid a heavy price as a result. For years, the U.S. auto companies ignored the growing demand for small cars. Japanese companies were attuned to the changing market, though, and they quickly stole market share from their American rivals.

The story is similar in the semiconductor industry. In the early 1960s, Fairchild, Philco, and General Electric were dominant forces in the industry. None of them recognized the growing importance of integrated circuits, however, and none of them is a major factor in the industry today. The process goes on and on. Five years ago, semiconductor companies felt pretty secure. They believed the capital intensity of semiconductor manufacturing would prevent new companies from entering the business. But dozens of new companies have been formed since then, largely because of new technologies.

The computer industry is another example. The major computer companies all ignored personal computers in the 1970s. Small startups began selling personal computers in 1976, but big companies like DEC didn't react until five years later. A major revolution took place right under the noses of the industry giants. Many large computer companies, such as Honeywell and Burroughs, will never fully recover.

Change has become a part of our lives, with one thing inexorably replacing another. We destroy the old and create the new.

In all industries, change is a tough competitor. What can marketing managers do to cope with this competitor? Two things.

First, marketing managers must constantly question their assumptions. They must ask questions such as: "What am I assuming about the market?" "What am I assuming about the competition?" "What things must happen to make my assumptions valid?" "Under what conditions are my assumptions no longer valid?"

Second, marketing managers must keep their ears to the ground. They must sense change as it is occurring. They must monitor the market, live with it, work with it. Oftentimes, changes do not show up in the numbers and statistics until it is too late. Marketeers must develop an intuitive sense of the market. They must work with customers and listen to them. They must meet with dealers and listen to them. And they must *really* listen. That is the only way they will spot changes in time to adjust.

Competitor 2: Resistance to Change

Sometimes companies recognize change is occurring in the marketplace, but they still don't react. For these companies, the competitor is resistant to change. Resisting change can be just as damaging as being oblivious to change. In either case, the company can get left in the technological dust.

Examples of resistance to change abound. Consider the case of Gary Boone. In 1972, as a young engineer at Texas Instruments, Boone came up with the idea for a full computer on a chip, later to be called the microprocessor. Boone got a patent on his invention, but he had trouble getting his colleagues interested in his work. He went around TI trying to sell the concept, but he was shot down everywhere. Other people looked on him as a young guy with a crazy idea.

Finally, Boone made enough noise to get a meeting with TI's top "guru" on computers. Boone went into his office, sat in front of the expert, and explained his idea for a computer on a chip. The expert looked at him with a condescending smile.

"Young man," he said, "don't you realize that computers are getting bigger, not smaller?"

There are similar stories with personal computers. Steve Jobs and Steve Wozniak tried to sell the idea of personal computers to their bosses at Atari and Hewlett-Packard. But their bosses weren't interested. So Jobs and Wozniak started Apple Computer. Intel also had a chance to get in on personal computers early. Several Intel marketing pros went to visit one of the early designers of personal computers sometime in the mid-1970s. They came back and reported: "Bunch of hobbyists. It will never be anything of a market."

Such resistance to change can destroy companies. Take a look at the following American industries: autos, steel, consumer electronics, calculators, machine tools, and textiles. In the mid-1960s, imports accounted for less than 10 percent of sales in the U.S. market in each of these industries. But American companies in these industries became resistant to change, while their foreign competitors did not. The result: In 1981, the U.S. imported 26 percent of its cars, 17 percent of its steel, 60 pecent of its consumer electronics (television, stereos, videocassette recorders), 41 percent of its calculators, 53 percent ot its machine tools, and 35 percent of its textiles.

What makes companies resistant to change? Sometimes bureaucracy is to blame, sometimes it's just that people are scared and intimidated by new things. People tend to get wedded to ideas. They look toward the past, rather than toward the future. When people move to a new company or a new project, they bring their histories with them. This experience can be useful, but it also can cause problems. Marketing people often say things such as: "This is the way we did it at my old company." This is helpful sometimes, but every once in a while, they should say: "Let's experiment and try something new."

The resistant-to-change demon rarely haunts young entrepreneurial companies. Entrepreneurs thrive on innovation and change. They are always willing to experiment with new ideas and new technologies. Resistance to change is anathema to entrepreneurs.

As entrepreneurial companies grow, however, they become more and more resistant to change. They begin to think more about high-volume production. They invest in special-purpose equipment that commits them to manufacture certain products in a repetitive, predictable way. That locks them into certain products and technologies. They begin to ask questions such as: "How do I keep my factories going?" and "How do I keep selling at this rate every month?"

In short, the factory becomes the central focus of the company. The company begins to worry more about manufacturing and less about serving the needs of the market. As a result, the company takes on the personality of a large company and becomes less likely to develop innovative new products. Small companies grab the lead in innovation.

The scenario is repeated time and again. The semiconductor memory business provides one example. Intel developed the first semiconductor memory chips, the 1K RAM. It clearly established itself as the leader in this new product category. But when Intel began working on the next generation of memory chips, the 4K RAM, it lost its innovative edge. Intel was committed to the approach it used in its money-making 1K RAM, but other approaches were better suited to the new generation. A small company called Mostek developed a more innovative 4K RAM, and it emerged as the new leader.

In many ways, mass production inhibits change and innovation. It is based on stability and predictability. When a company moves into mass production, it becomes resistant to change. It wants to build up economies of scale. Innovation can disrupt that effort.

Growth companies face a difficult challenge. They must figure out a way to move toward high-volume production, while not losing the innovative spirit that made them successful in the first place. They must continue to use innovation and change as allies, not competitors.

Competitor 3: Public Knowledge About the Product

An uninformed customer is easily satisfied. But there aren't many uninformed customers around these days. Customers today have access to more product information than ever before, and they study it carefully. With technology products, customers are becoming more "technology-literate."

Technology literacy is clearly a challenge for manufacturers. Customers are no longer pushovers. They want to understand more. They are skeptical and critical, and more often dissatisfied. Manufacturers must meet a higher level of expectations.

Consider the amount of computer information reaching the public these days. A few years ago, there were a handful of computer magazines. Now there are hundreds. A few years ago, *Time* and *Newsweek* hardly ever wrote about computers. Now, they both have computer editors. A few years ago, television news never ran stories about computers. But for the introduction of Macintosh, all three networks ran stories, as did more than twenty individual stations.

As the quantity of coverage has increased, the quality of coverage has improved. Journalists themselves are becoming more technology-literate. Until a short time ago, computer companies could use journalists to spread just about any message they wished. The journalists didn't know enough about technology to critically evaluate computer companies and their products. That has changed. Many journalists use personal computers, and are quite knowledgeable about them. When a company introduces a new computer today, journalists want to evaluate the computer themselves. They won't take the company's word about what the machine can and cannot do. In effect, the journalist becomes an evaluator for the public.

To succeed, companies must turn this increasing knowledge of their products from an obstacle into an asset. Rather than battle against a skeptical and critical public, they should learn from it. They should elicit feedback from customers, then adjust their products and strategies to meet the market needs.

Some consumer-goods companies are already quite successful at putting customer dissatisfaction to their advantage. According to a *Wall Street Journal* article, Procter & Gamble phones or visits 1.5 milliion people each year to ask about P&G products. P&G researchers ask hundreds of detailed questions to find out why customers are dissatisfied and what actions P&G should take to improve its products.

The same article quoted from a recent study by the U.S. Office for Consumer Affairs: "Many managers view complaints as a nuisance that wastes valuable corporate resources. However, the survey data suggest that complaints may instead be a valuable marketing asset. Responsive companies were rewarded by the greatest degree of brand loyalty."

Technology-based companies should learn a lesson from this. As customers become more knowledgeable—and more critical—about technological products, companies must become more sensitive to customer needs. The philosopher John Stuart Mill once said: "Better to be Aristotle dissatisfied than a fool fully satisfied." Customers of technological products are taking Mill's advice, and companies must adjust.

Competitor 4: The Customer's Mind

People in technology-based businesses tend to think decision-making is a simple and rational process. They are wrong. Indeed, when a customer considers buying a product, the decision-making process is neither simple nor rational. All types of fears, doubts, and other psychological factors come into play.

Winning over the customer's mind is the central challenge of marketing. The customer's mind can be seen either as a competitor or as a competitive tool. Sometimes the customer's mind can act as an obstacle to success. But if companies can understand the customer's mind, they can use psychological factors to their advantage.

All types of things influence the customer's mind. Indeed, the battle for sales is largely a psychological battle. As I explained before, decisions are made largely on intangible factors such as quality image, support, and leadership. In *Future Shock*, Alvin Toffler describes the psychological battle this way:

For even when they are otherwise identical, there are likely to be marked psychological differences between one product and another. Advertisers strive to stamp each product with its own distinct image. These images are functional. The need is psychological, however, rather

> *than utilitarian in the ordinary sense. Thus, we find that the term "quality" increasingly refers to the ambiance, the status associations—in effect, the psychological connotations of the product.*

Customer attitudes toward a product are not developed by a single event or a single advertisement. Rather, customer attitudes develop gradually. They are constantly changing and evolving throughout the decision-making process—and continue to evolve after the decision is made. The "product image" is simply the accumulation of all these attitudes.

The customer's mind can be influenced at every step during the decision-making process. First, the people become aware of the existence of the product. Then, they recognize the need for the product. At that point, they will try to find out more about the product. They might talk to people who already have used the product, or read reviews written by experts. They might use the product on a trial basis. At each stage, their attitudes are modified and reformed. After the purchase, customers' attitudes continue to evolve as they use the product. They expect a certain level of product support and product performance. If their support and performance fall short of expectations, their attitudes will turn negative.

Throughout the entire process, "psychological bogeymen" affect the customer's mind. These bogeymen include all types of doubts and fears that surround the product, making the customer uneasy about making the purchase. Customers might worry about such things as:

- Is the company going to be around for a long time?

- Am I going to be able to get product support after the purchase?

- Will the manufacturer be able to supply future generations of products?

- Will I be technologically behind if I buy this company's products now, rather than waiting for its competitor's upcoming product?

In winning the battle for the customer's mind, companies might fight against these psychological bogeymen. They must provide "comfort factors" that put the customer's mind at ease.

For complex technical products, these comfort factors are particularly important. A company must convince customers that it is financially and technically strong enough to meet all of the customer's future needs.

At the same time, companies can try to influence customer attitudes toward competitors' products. With its FUD strategy mentioned earlier, IBM works from both sides. It adds comfort to its own products and psychological bogeymen to its competitors' products. Clearly, the strategy has been quite successful.

To succeed in the market, companies also must work to understand the customer's mind. It is not enough to know what competitive products are on the market and who is using them. Marketeers must understand the psychological bogeymen and comfort factors that influence the customer's mind, then use these psychological factors to their advantage.

Competitor 5: The Commodity Mentality

What is good for manufacturing is not always good for marketing.

For efficient, low-cost manufacturing, nothing beats commodities. By churning out the same commodity product time after time, manufacturers can work all of the kinks out of the production process. As volume increases, manufacturers move down the so-called learning curve, and their costs drop lower and lower.

But a marketing strategy that depends on a commodity mentality can be deadly. Customers usually prefer custom-made, "just-for-me" products. They want their needs satisfied exactly. We are in an age of diversity, and people want to feel as if they are getting something special.

Companies that view their products as commodities will have an increasingly difficult time competing, especially in evolving markets. Companies that sell commodity products can attract customers only by keeping prices low. Competition generally degenerates into a struggle for price leadership, and no one ends up making much money.

How can companies get out of this commodity trap? Meshing the differing needs of manufacturing and marketing isn't always easy, but it can be done. The trick is to view products as more than physical entities. Even if a company manufactures commodity-like products, it can differentiate the products through service, support, or target marketing. It can leave its commodity mentality in the factory, while bringing a mentality of diversity to the marketplace.

To move away from the commodity mentality, companies must view their products as problem-solvers, and then sell the products on that basis. In his article "Marketing Success Through the Differentiation of Anything," Theodore Levitt describes the approach this way:

To the potential buyer, a product is a complex cluster of value satisfactions. The generic thing is not itself the product...A customer attaches value to a product in proportion to its perceived ability to help solve his problem or meet his needs.

An automobile, for example, is not just four wheels and an engine. It is a means of fulfilling customer needs, psychological and otherwise. Manufacturers can differentiate their automobiles according to the needs they fulfill. One can be positioned as a "status" product, another as a "performance" product, even if the products themselves are quite similar. If automobiles were marketed solely on the basis of their specifications (the number of cylinders, the size of the engine, and others), customers would perceive them all as being very much alike. Indeed, "specsmanship" marketing is a sure sign of a commodity mentality.

The personal computer provides another example. Everyone views the personal computer in a different light. Many managers see it as a productivity tool that provides increased freedom to information-users. But some MIS managers see the personal computer as a device that causes information and other resources to be used inefficiently within large organizations. The same product, but very different perceptions.

The perception of personal computers also changes with time. At first, the Apple II was seen as a hobbyist computer. Then a small-business computer. Then as a "vertical market" computer—that is, a computer able to serve many different, specialized applications. The Apple II has remained the same. But the marketplace has changed, and so has Apple's marketing strategy. Apple has manufactured the Apple II like a commodity. But in its marketing, Apple has tried to make the Apple II special to every customer. It has stayed away from a commodity mentality.

Competitor 6: The Bigness Mentality

Edward Schumacher was certainly right when he coined the phrase "Small is Beautiful." Just consider the following statistics:

- More than half of the innovations in the United States in the last twenty years have come from companies with fewer than 200 employees.

- A study by Massachusetts Institute of Technology professor David Birch showed that companies with fewer than twenty employees created 60 percent of all new jobs, and companies with fewer than 500 employees created 86 percent of all new jobs.

- Of the 9 million jobs created between 1966 and 1977, 6 million were created by small businesses, 3 million by government, and zero by Fortune 1000 companies.

- Small companies are more efficient with R&D. A study by the National Science Foundation showed that small companies (fewer than 1,000 employees) produced four times as many innovations per R&D dollar as medium-sized companies (1,000 to 10,000 employees) and twenty-four times as many innovations as large companies (more than 10,000 employees).

Indeed, study after study shows that small companies are more innovative and productive than larger companies. Unfortunately, as small companies grow and become large companies, most of them run into the same problems as other big companies. They become less creative and less dynamic. They begin to suffer from what I call bigness mentality.

A major element of bigness mentality is an aversion to risk. Small companies cannot afford to take the safe path. They could not compete with established companies on that basis. They must come up with new ideas, experiment with new approaches, try new things. They must innovate or they will not survive.

As companies grow, they become more reluctant to take risks. If the company decides to go public, as most do, it is evaluated by the financial community on a quarterly basis. If financial results slip one quarter, the stock price could plummet and the company could have trouble raising new funds. So public companies must play it safe. They can't afford to take short-term risks, even if they might pay off with long-term benefits. Wall Street thinks short term, not long term.

Corporate bureaucracy also reduces risk-taking and innovation. As small companies grow, they restructure themselves to look and act like big companies. Decisions are made by committees, not individuals. As a result, decisions tend to be compromises, not bold new approaches. People begin to worry more about avoiding mistakes than creating new ideas.

Take advertising decisions. Advertisements run by small companies tend to be much better than those run by big companies. They are more creative, more aggressive, more interesting, more attention-grabbing. Why? Big companies usually have large advertising departments that make decisions by committee. No one is willing to stick his neck out.

As a growing company adds new committees and new levels of bureaucracy, it is slower to notice new opportunities in the market and slower to respond to changes in the market. Its "corporate reaction time" shoots up. Earlier, I discussed Intel's quick reaction to the challenge from Motorola's 68000 microprocessor. Within seven days, Intel designed a new strategy, presented the plan to 150 managers from around the world, and began to put the plan into action. At most large companies, it probably would have taken seven days just to arrange the initial meetings.

How can companies avoid the bigness mentality? One way is to maintain small, entrepreneurial project groups within the company. IBM took this approach with its personal computer. The company created an independent group in Boca Raton, Florida, and gave the group an unusual degree of freedom. In doing so, IBM acted in an un-IBM way. Although IBM was still a big company, it was thinking like a small one. It broke all its own rules, and took some risks. The risks certainly paid off.

Another key is to avoid compartmentalization in the corporate organization. Many growing companies break various functional groups into different divisions, then make it difficult for those groups to interact. In small companies, people in engineering, marketing, and sales talk regularly and exchange ideas. This interaction is vital to creativity and innovation, but it is usually missing in large companies.

In his book *The Next American Frontier*, Harvard professor Robert Reich argues that large companies must develop new forms of organization which allow greater interaction among different groups. He writes:

---

*P*recision manufacturing, custom-tailoring, and technology-driven products have a great deal in common. They all depend on the sophisticated skills of their employees, skills that are often developed within teams. And they all require that traditionally separate business functions (design, engineering, marketing, and sales) be merged into a highly integrated system that can respond quickly to new opportunity. In short, they are premised on flexible systems of production.

Only with this type of flexibility can companies avoid the bigness mentality and maintain their creativity and productivity as they grow.

Competitor 7: Broken Chains

The business world is full of chains and connections. Processes and products are linked to one another in a great chain that ultimately connects companies and customers. No problem or business decision is isolated or self-contained. Companies get into trouble when they think about one link at a time, focusing on advertising or public relations or manufacturing, without recognizing that all these functions are interrelated. By ignoring the linkages, companies end up with a broken chain— and a failed product.

To get more specific, consider the most important chain: the product-customer chain. This chain connects everything in the product development and marketing process. It starts with the design and planning of the product. Other links include product development, manufacturing, marketing, sales, distribution, product support, and service. The final link is the customer.

All of these links are part of one common process with one common goal: serving the customer. What a company does in one stage of the process can affect many other stages. Manufacturing affects marketing, and marketing affects sales. If any link in the chain is broken, the primary goal of the chain—serving the customer—goes unfulfilled.

A chain is only as strong as its weakest link, so companies must pay attention to every link. They also must maintain strong connections between the links. Different departments must talk to one another and work with one another. If a company fragments into a bunch of loosely connected fiefdoms, it will lose to a more coordinated competitor. The problems at Xerox's Palo Alto Research Center, discussed earlier, provide an example. The researchers in Palo Alto were top-notch, but they rarely talked to other groups within Xerox. Thus, the product-customer chain was broken, and products got to the market late, if ever.

Another important marketing chain involves what are known as "consumption patterns." These patterns are, in fact, product-product chains: They link together the sales of different products. When people buy lettuce in a supermarket, for instance, they also buy an average of $2 to $3 of complementary products, such as salad dressing and croutons. By understanding how this chain works, a supermarket might put its lettuce on sale in order to spur sales of the complementary items.

The same type of chain works in the computer business. People that buy personal computers buy a variety of complementary products, including software, printers, and modems.

These products are not isolated products. Each one helps sell the others.

When consumption-pattern chains are broken, trouble is sure to follow. If one link of the chain—say, software—is missing, sales of all other products will slump. Intel's success in the microprocessor business is largely because of its understanding of consumption patterns. Intel sells not only the microprocessors themselves, but also the peripheral chips and development systems needed to put the microprocessors to use. Intel constantly adds new types of peripheral chips and microprocessors, and each new product enhances the sales of the others.

Distribution strategy also plays an important role in product-product chains. In some cases, all the products in the consumption-pattern chain are available, but they are sold through different distribution channels. That can be just as bad as a missing link. The pieces all exist, but they are not linked together into a strong chain. The customer cannot easily buy everything he needs, so he might end up buying nothing. For this reason, retailers usually like to handle full product lines, not just individual hot products.

Another important chain is the chain between different markets. Sales of a product into one market influence sales of the same product into other markets. In the personal-computer business, for instance, the home and office markets are strongly linked. People who use personal computers at work are more likely to buy them for their homes. The reverse also holds: People with computers at home push for greater use of computers in the office. In many cases, parents will learn from their children. They buy home computers for their children, then they, too, become interested in the machines. Before long, they want to use computers in their businesses.

The education market is another link in this chain. Children who use computers in schools often pressure their parents into buying home computers—usually the same brand used in the school. The university market is also important. Today's college students are tomorrow's decision-makers in the business world. In a few years, they could be deciding what types of computers to buy for their businesses. They are likely to buy the same brand of computer they used in college.

Computer companies are now scrambling to take advantage of this linkage, offering computer systems at great discount to schools at all levels. IBM, for instance, has donated thousands of computers to elementary schools, high schools, and colleges. Apple has donated more than 10,000 Apple II computers to schools in California, and it sells its Macintosh computer to select universities at less than 50 percent of the retail price. As long as the home-office-school chain remains unbroken, these giveaways could lead to big sales down the road.

Competitor 8: The Product Concept

What do IBM, AT&T, CBS, Dow Jones, and Apple have in common? Five years ago, the answer would have been not much. IBM sold big computers and office equipment. AT&T was in the telephone business. CBS was a television network. Dow Jones was a publishing company. And Apple sold personal computers.

Today, however, all five companies compete against one another, at least indirectly. All are involved in the information business. All offer equipment and services that enable customers to access information more quickly and efficiently. In the future, they will compete directly with one another more and more often.

In this type of environment, companies cannot afford to think about their products too narrowly. They must look for opportunities—and expect competition—in every possible direction. A company with a narrow product concept will move through the market with blinders on, and it is sure to run into trouble. The product concept itself will become a competitor.

Earlier, I mentioned the classic business school example of the railroads. Had the railroads considered themselves "transportation companies," rather than railroad companies, they might have moved into the airline business. Instead, the railroads stuck to their narrow product concept and watched the new airline companies steal much of their business.

The same situation exists now in many evolving industries. Dow Jones, for instance, cannot afford to think of itself as a magazine and newspaper company. It must see itself as an information company. It must provide information in whatever form customers desire, whether it is written on paper, broadcast to radios, or sent over telephone lines to computer screens.

Similarly, a personal-computer company should not view its product simply as a box with a keyboard and a display. If it sees its product that way, the company will have a narrow view of its competition. It will see other personal-computer companies as its only real competitors, and it will plan its strategies with a false sense of the market.

In fact, many different products could compete with personal computers. Application-specific devices, such as pocket pagers and stock-quotation devices, are potential competitors. So are computer terminals and touch-tone telephones. Many companies are setting up information networks that allow users to access information through inexpensive "dumb" terminals. If people use these networks often, they might just buy a terminal, not a personal computer.

Home televisions also could be competitors. In two-way cable-television systems, subscribers can use their televisions for all types of information services. They can order airline and theater tickets, check their bank account balances, pay their utility bills, and check the prices of the stocks they own. Televisions will become even more formidable competitors to personal computers as manufacturers begin to build computers right into the sets that they sell.

In developing their marketing and positioning strategies, personal-computer companies must consider all these new competitors, and try to anticipate other challengers. If they limit their product concept and keep their blinders on, they are sure to be blind-sided in the market.

Competitor 9: Things That Go Bump in the Night

No matter how well a company understands its market, it is bound to be taken by surprise sometimes. New technologies, new companies, new applications all can shake up an industry in a hurry. I call these unanticipated events "things that go bump in the night." Companies don't see them coming. But like the iceberg that sank the *Titanic*, they can do a lot of damage.

There are more things going bump in the night today than ever before. The prime reason is the speed-up in technological innovation. According to one estimate, 99 percent of all technological innovations in the history of mankind have occurred in the past twenty years. Every year, there are more innovations than the year before, and each innovation holds the potential for shaking up a company, if not an industry.

The base of scientific knowledge, from which technology evolves, is continuing to grow rapidly. According to some estimates, more than 90 percent of all the scientists that ever lived are alive today. More important, scientific knowledge is being put to use more quickly than ever before. Engineers are constantly shrinking the time it takes to translate scientific advances into new technological products. One researcher, studying twenty major innovations, found that the time lag between scientific discovery and technological product has dropped by 60 percent since 1900.

No company in a technology-based industry is safe from unanticipated bumps in the night. The steel industry, the petroleum industry, even the textile industry—all have been jolted by technological change. A decade ago, the major pharmaceutical companies thought their industry was fairly mature and stable. Then came the development of recombinant DNA technology, and now dozens of new companies are challenging the products of established industry leaders.

The semiconductor industry has been predicting a major shakeout for fifteen years now. In the late 1960s, many industry experts predicted the semiconductor industry would soon resemble the auto industry, with only three or four leading manufacturers. They argued the business was too capital-intensive for new companies to join. At the time, there were about ninety-five semiconductor companies. Today, there are more than 200.

Steve Jobs of Apple made a similar prediction. He said the personal-computer industry was too capital-intensive to support new entrants, and he predicted a major shakeout in the industry. Certainly, there will be some type of shakeout, but there is no way the industry will consolidate into a handful of companies, at least not in the near future. There is still plenty of room for technological innovation, and that means plenty of room for new competitors.

There is no way for companies to avoid bumps in the night. But companies can be prepared for them. They can stay humble, expect the unexpected, and react quickly when the unexpected occurs. They must understand that no company is too big, and no industry is too capital-intensive, to be shielded from the havoc caused by technological innovation.

Competitor 10: Yourself

This is the toughest competitor of all. Machines don't compete, people do. People have to spot the market opportunities and take advantage of them.

There are many ways people end up competing with themselves. When people underestimate their own ideas, just because the ideas have never been tried out before, they are competing with themselves. When, on the other hand, people develop an air of omnipotence and believe they can't fail, they also are competing with themselves. When people are unwilling to listen, when they are unwilling to change, when they are unwilling to experiment—in all these cases, they are competing with themselves.

People must leave themselves open to think creatively. With markets changing so rapidly, managers must be able to analyze new situations and apply creative approaches. Old approaches to new problems simply won't work.

Above all, managers must pay attention to the market. They must listen and respond to it. They must not underestimate their competition—or overestimate it. And they must continue to experiment.

If managers adopt this pattern of thinking, this frame of mind, they can avoid the biggest problem of all: turning themselves into a competitor.

*I*nsanely great!

Steve Jobs

8 Putting Ideas to Work: Marketing Macintosh

Macmarketing

Annual meetings for Fortune 500 companies are usually rather boring affairs. The corporate secretary announces the predictable results of proxy votes. Other executives read off long lists of corporate accomplishments and financial results. If the corporation has performed well in the past year, the stockholders applaud politely. If not, they ask a few questions and worry about their dividend checks.

But when Apple Computer stockholders met on January 24, 1984, it was hardly an ordinary annual meeting. More than 2,500 people jammed into the Flint Center in Cupertino, California. The atmosphere was part carnival, part revival meeting. Like a minister at the pulpit, Apple chairman Steve Jobs preached to the gathered masses. He told them that January 24 marked the beginning of a new revolution in personal computing. The cheers from the crowd rose to a crescendo as Jobs walked over to a table carrying a suitcase-sized case. Jobs unzipped the case, and the world got its first look at the Macintosh computer. The crowd screamed its approval.

Apple backed up the Macintosh introduction with an enormous media campaign. It ran twenty-four Macintosh commercials during the Winter Olympics, and it paid for a twenty-page advertising insert to appear in eight magazines. To build up media interest, Apple sent Macintosh press kits, including Macintosh T-shirts, to 3,000 financial analysts and journalists. Articles about Macintosh seemed to be everywhere. The computer appeared on the cover of thirteen magazines. Television news shows ran features about Apple and its revolutionary new product. So did *Time* and *Newsweek*. Computer dealers were besieged with inquiries and orders. Within 100 days, Apple sold more than 75,000 Macintoshes.

To the outside world, the marketing of Macintosh seemed like a big promotional event. But success does not come that easily. Hidden behind the splashy introduction were hard years of marketing development and planning.

The story of the Macintosh success presents an instructive example of how to market new products in an era of rapid change. It is a story of a small group of people who succeeded in positioning a new product while ignoring many traditional rules of marketing. The members of the Macintosh team relied to an extraordinary degree on their collective intuition about the market, performing hardly any statistical analyses. They succeeded by constantly refining and readjusting the positioning of their product to match the ever-changing market environment.

In this chapter, I'll present an inside look at the Macintosh marketing effort, relying heavily on the insights I gained while consulting for the Macintosh team. The chapter is, in many ways, an appropriate conclusion to the book: It will show how Apple weaved together many of the dynamic-marketing strategies described in this book, and ended up with a big success.

Macbeginnings

The story of Macintosh begins nearly five years before the product's introduction. Jeff Raskin, an engineer who had written technical publications and manuals for Apple, came up with the initial idea for Macintosh in 1979. While many engineers were trying to modify or enhance the enormously successful Apple II computer, Raskin believed Apple needed a radical new approach to computing.

The Apple II was itself a pretty radical product in the world of computing. Introduced in 1976, the Apple II created an entirely new product category called personal computers. By 1980, Apple held an 80 percent share of the personal-computer market. It had changed forever the way people thought about computers. Computers were now light enough to carry and cheap enough to sell through retail stores. Across the country, thousands of software developers dedicated themselves to designing programs for the Apple II, opening up myriad new applications for personal computers in homes, offices, and factories.

But Raskin's intuition told him the technologies used in the Apple II would never bring computing to the masses. The Apple II was fine for hobbyists and people willing to spend long hours learning arcane computer commands. But Raskin wanted to build a computer that anyone could use. At first, he called it "Everyman's Computer." Raskin knew that the Apple II and its direct descendents would never fill that role. Raskin didn't do any market research or customer surveys; he relied on intuition. He knew that Everyman's Computer would have to be far easier to use than the Apple II or anything else then on the drawing boards.

To make his computer easier to use, Raskin relied on a new "user interface"—that is, a new way for users to interact with the machine. Communication with the machine was based on pictures more than words. Using a hand-held device called a mouse, users could point to pictures that appeared on the screen. The result was a much more natural interaction. Rather than typing out strange commands, as they had to do with other computers, Macintosh users would simply point. The screen, for instance, could simulate a desktop. Users could point at items on the screen just as they would pick up pieces of paper on their desks.

The original designs for Macintosh were quite different from the final product. Raskin and his colleagues figured the machine would weigh less than ten pounds. In fact, it ended up weighing seventeen pounds. They hoped that the machine could be priced under $1,000. Instead, it came to market at $2,495. The original design was battery-operated. The final product was not. One thing about the Macintosh project never changed, though: the commitment to making Macintosh easy to learn and use.

Raskin's dream computer probably would have stayed just a dream had it not been for Steve Jobs. When Jobs saw Raskin's design, he fell in love with it. Like Raskin, Jobs believed Apple had to try some radical new approaches to computing. Jobs had tried to get involved in Apple's Lisa project, but was not really welcomed there. He was a founder and a top officer of Apple, but in many ways Jobs was a man without a company. Macintosh captured his imagination. He became head of the Macintosh project.

Under Jobs, the Mac project became a company within a company. In effect, Jobs was heading a startup company, just as he had done in the early days of the Apple II. The Mac group remained rather small. In early 1982, there were still only twenty-five members.

In the rest of Apple, the Mac project had low credibility. Few people expected it to succeed. But inside the Mac project, everyone was a believer. Mac team members shared a vision of what personal computers could and should be. They developed an extraordinary sense of community and comraderie. They often would sit on the floor and talk through the night. Their intuitions were strong: They knew they had a winner.

Macaudience

For the first two years of the Macintosh project, there was no marketing staff. The project was driven purely by technology. The Macintosh team and its budget were both small, and team members wanted to put all their time and money into development, not marketing.

In 1982, though, the Macintosh group began adding some marketing people and holding strategy sessions. Four or five people (including myself) met every two weeks to talk about positioning for Mac. The meetings were mostly brainstorming sessions. We spent a great deal of time trying to figure out how to "create" the market and what users to target, who was going to use Macintosh, and how we could best communicate new ideas to those potential users.

We decided that Mac's target audience would not be a traditional market segment. For some products, you can take a neat slice of the market—for example, all small businesses in service industries, with annual sales between $500,000 and $1 million. That type of segmentation wouldn't work for Mac. Mac cut across the usual boundaries. We needed to come up with what I call a "concept market." Most marketing managers look to divide a market along demographic or geographic lines. But concept markets are divided along "psychographic" lines. That is, they include people with similar attitudes and beliefs.

We came up with the idea of "knowledge workers." These are people who typically sit at a desk during the day. They create ideas, make plans, analyze data. Knowledge workers exist in many different settings. Some work in large offices as professionals, others in homes as consultants, and still others in dormitories as college students. One internal Apple marketing plan described knowledge workers this way:

> *K nowledge workers are professionally trained individuals who are paid to process information and ideas into plans, reports, analyses, memos and budgets. They generally sit at desks. They generally do the same generic problem solving work irrespective of age, industry, company size, or geographic location. Some have limited computer experience—perhaps an introductory programming class in college—but most are computer naive. Their use of a personal computer will not be of the intense eight-hour-per-day-on-the-keyboard variety. Rather they bounce from one activity to another; from meeting to phone call; from memo to budgets; from mail to meeting. Like the telephone, their personal computer must be extremely powerful yet extremely easy to use.*

After some rough calculations, Apple figured there were about 25 million knowledge workers in the United States that might use Macintosh computers. That included 5 million in small businesses (less than $5 million in annual sales), 5 million in large businesses (Fortune 2000 companies), and nearly 9 million in medium-sized businesses. The other knowledge workers were in the college and home markets.

At the time, not many knowledge workers were using personal computers. People need a significant amount of expertise and training to use the Apple II. To really appreciate the machine, you have to take the top off and put new boards in. Selling the Apple II was like selling a telephone you had to put together yourself. But Mac would be different. It would be nonintimidating and easy to use. People could learn to use it in two to four hours, rather than the twenty to forty it took for the Apple II. Knowledge workers would feel comfortable with it. With Mac, Apple could tap into a new market of 25 million workers. Mac could become the standard product of knowledge workers.

We began to view knowledge workers as the next stage in the adaptation sequence for personal computers. As discussed in Chapter 4, most new products are accepted by the market in stages: First by Innovators, then by Early Adapters and Late Adapters, finally by Laggards. Before Macintosh, only Innovators had purchased personal computers. They were willing to read a 400-page user's manual and spend tweny to forty hours learning how to use the computer. The computer was an important part of their lives. Some were known as "spreadsheet junkies."

But the market had begun to run out of Innovators. In the same way many Americans had waited for Interstate 80 before heading West, so too were many people waiting for easier-to-use computers. Most knowledge workers were part of this group. The next generation of computers had to address the needs of these workers.

We spent hours discussing such questions as: Who are knowledge workers? Where are they? How can we identify them? The strategy was to make Mac so unique and innovative that knowledge workers would be romanced into using it.

At the marketing meetings, we also spent a lot of time thinking about words and language. How could we communicate about Mac? We decided we had to create our own vocabulary. Revolutions create their own language. We knew language would be very important in positioning Mac. We knew Mac was radically different from other personal computers on the market, and we didn't want them compared. We didn't want people to compare operating system v. operating system, keyboard v. keyboard. Instead, we would create our own words. Then, when the press and dealers talked about computers, Mac would stand out. It was a type of forced differentiation.

Some words were obvious, like "mouse" and "user interface." But most of the discussion focused on the word "appliance." Mike Murray, the marketing manager for Mac, felt that Apple could position Mac as an information appliance. Some of us, myself included, were somewhat skeptical. We argued that computers were more complex than refrigerators and other traditional appliances. We also worried that the word "appliance" would target Mac toward the home, not the office.

But Murray insisted. He gave away miniature food processors to members of the Mac team. In August 1982, when Murray wrote the first product plan for Mac, he drew heavily on the appliance metaphor. He wrote:

> *Think of Mac as an appliance. A thing applied as a means to an end. Like a Cuisinart in the kitchen, one could live without a Macintosh on the desk. Yet the increased personal productivity combined with the opportunity for personal creative expression will hit hard at our customers' psychic drives. Perhaps only 15-20 percent of a person's working time will be spent using a Mac, but as with the Cuisinart, it will make all the difference in the world. Our customer will find it very difficult to return to the "old way" of doing things. Macintosh will become an integral part of life at a desk. In fact we would like to see the day when a freshly hired product manager for a Fortune 500 company walks up to his new desk and finds a telephone, a couple pens, a tablet of paper, a company magazine and a Mac.*

Macconflicts

Differentiating Mac from competitors' products was only part of the marketing problem. Just as important, the Mac marketeers had to figure out a way to differentiate Mac from the rest of the Apple product line. All Apple computers, those on the market and those in development, aimed at roughly the same audience. There were serious fears that each Apple computer would cannibalize the others.

All companies with more than one product face these internal conflicts. The typical solution is to divide up the turf. Target one product at small businesses, another at professionals. Sell one product at a premium price, another at a discount. If products overlap too much, dealers and customers will be confused. Dealers won't know how to sell the products, and customers won't know which one to buy. Each product must give up a bit of its territory for the sake of clarity.

At Apple, however, this normal approach seemed impossible. Each Apple product group was an individual fiefdom, and each was more interested in competition than cooperation. Many Apple products were targeted at small businesses and professionals, but the different Apple marketing groups wouldn't sit down together to resolve the conflicts.

Perhaps the biggest conflict came between Mac and Lisa. The two projects had very different roots. The Lisa project was a major corporate effort, with lots of people and lots of money, while the Mac project was more of an in-house guerrilla group. By 1982, the two projects were on a collision course.

Lisa's development group hoped to establish its product as a new standard in personal computing. Mac's group had the same hopes. Lisa's team planned to make its computer dramatically easier to use than traditional computers through the use of a new graphics-based user interface. Mac's team had the same plans. Lisa used the Motorola 68000 microprocessor. Mac used the same. Lisa used a mouse as a pointing device. So did Mac.

Even the target audiences were growing more alike. In the beginning, Mac's designers shunned business applications. They wanted to bring computer power to the "masses," not money-hungry businessmen. They were suspicious of anyone who read *The Wall Street Journal* and actually enjoyed it. But when Jobs took over the Mac effort, he steered it toward office applications and knowledge workers—aiming at much the same audience as Lisa.

The two groups became intensely competitive. Each wanted to outshine the other. The general managers of the two groups, Jobs and John Couch, even made a personal bet over which computer would reach the market first. The loser would have to throw a celebration party for the winner.

Conflicts also existed between Mac and the aging Apple II. Most people saw the Mac as the replacement for the Apple II. It would make the Apple II obsolete. That could cause big problems for Apple, however. Despite its incredible growth, Apple was still, to a large extent, a one-product company. It was doing $500 million in sales with one basic product, the Apple II. What would happen if that one product became obsolete overnight?

When developing a successor to an existing product, companies usually try to arrange for one product to gradually replace the other. As sales of the new product shoot upward, sales of the old one gradually fall. Many Apple marketeers worried the transition from the Apple II to the Mac would not be so smooth. When Mac was introduced, they figured, Apple II sales would plummet. But Apple could only build up its production of Macs slowly. The transition could be disastrous.

Mactrouble

The year 1983 was one of turmoil and changes for Apple Computer. It started smoothly enough. In January, Apple introduced its Lisa computer. (Lisa had moved through the development cycle faster than Mac, allowing John Couch to collect on his bet with Jobs.) The marketplace quickly acknowledged Lisa as a revolutionary product. Magazines ran glowing reviews about the new machine. Some people complained about Lisa's $10,000 price tag, but nearly everyone agreed Lisa set a new direction for the personal-computer industry.

The enthusiastic reception for Lisa was due, in part, to a remarkably favorable market environment. Everything seemed to be going Apple's way. The personal-computer industry as a

whole was still expanding rapidly. Almost everyone was pros-
pering—Atari, Osborne, Commodore. *Time* magazine had even
selected the personal computer as its "Machine of the Year" for
1982.

The entry of IBM into the personal-computer market in 1981
had scared some competitors, but IBM's presence seemed to be
helping rather than hurting other manufacturers. IBM stamp of
approval gave added credibility to the new industry. The rising
tide of enthusiasm about personal computers lifted all boats in
the industry.

No one benefited from the booming market as much as Apple.
Apple was still the clear industry leader in sales and profits. Just
five years after Steve Jobs and Steve Wozniak started the company in a garage, Apple was a member of the elite Fortune 500.
Never before had a company joined the ranks of the Fortune
500 so quickly.

But as it so often does in dynamic industries, the market
environment for personal computers shifted dramatically in the
first six months of 1983. Prices continued to drop and competition continued to grow, putting the squeeze on industry profits.
The personal-computer industry was no longer a paradise. One
company after another ran into financial trouble. Osborne and
Victor filed for bankruptcy. Texas Instruments and Atari lost
hundreds of millions of dollars.

Apple, too, began to struggle. Its quarterly profits fell for the
first time in its history. Even worse, the heralded Lisa computer
faltered in the marketplace. Production problems forced Apple
to delay shipping the new computer, and Lisa never regained its
initial momentum.

Apple had expected to sell lots of Lisas to knowledge workers
at large corporations, but the shifting environment foiled Apple's
plans. Before 1983, personal computers were seen as stand-
alone machines. Individuals in large corporations bought personal computers without even consulting the manager in charge
of management information systems. Apple hoped to sell Lisas
the same way. It figured people would be willing to spend
$10,000 for a technologically advanced stand-alone system.

Things didn't work that way. MIS managers began to worry
about the uncontrollable influx of personal computers. Many
corporations were getting stuck with a random collection of

machines from many different manufacturers. In most cases, these machines were incompatible. They couldn't use the same software or peripherals, and they couldn't be connected into networks to share data and information. MIS managers declared war against these unauthorized computers. They sought to regain control over all computer resources in the company.

That was bad news for Apple and good news for IBM. MIS managers had bought from IBM for years, and they trusted IBM. They began to buy IBM personal computers by the hundreds, even thousands. Apple couldn't get its foot in the corporate door. MIS managers saw Lisa as too expensive and lacking in networking and data communications capabilities. And the Apple II simply wasn't powerful enough for many new applications.

IBM continued to gain momentum. Its personal-computer market share grew from 18.4 percent in 1982 to 30 percent in 1983. IBM seemed invincible. Rumors circulated about IBM's next personal computer, an inexpensive home computer code-named Peanut. Analysts and retailers predicted Peanut would be the first true mass-market computer. They declared Peanut a winner even before it was introduced. Stock prices of other personal computer companies fell sharply in anticipation of Peanut. Apple's Macintosh seemed out of step with the environment. Apple was getting ready to introduce a $2,500 computer while everybody in the industry was talking about the $500 Peanut.

In the span of a few months, IBM's image in the industry had changed dramatically. To customers and analysts, IBM was the dominant industry leader. To other manufacturers, IBM was now Enemy No. 1. When IBM introduced its PC, Apple had run an ad in *The Wall Street Journal* saying "Welcome IBM." But now, Apple wanted to take in the welcome mat. A joke circulated around Apple's Cupertino headquarters:

Question: What are the two biggest lies in the world?

Answer: The check is in the mail and Welcome IBM.

Maccomeback

When John Sculley joined Apple as chief executive in April 1983, he faced a desk full of troubles. With the disappointing sales of Lisa, Macintosh was now more important than ever. It was no overstatement to say the future of Apple depended on Macintosh.

Apple managers were still enthusiastic about Macintosh, but they were depressed about the market. The environment was nasty. No one knew how the market would accept Macintosh, and no one was quite sure how to market Mac. The marketing for Mac had been kicked around for a few years as the marketplace went up and down, back and forth, constantly changing. Somehow, Apple had to fit Macintosh into the current environment.

Sculley also had to clean up two other problems. First, he had to make sure the company remained profitable. Apple had built up for huge Lisa sales that never came. The company was overextended and profits were falling. If profits ever dipped into the red, it would be disastrous for Apple. The company would lose its credibility and its corporate positioning. In people's minds, Apple would be thrown in with Osborne, Atari, and Texas Instruments. People would start questioning Apple's future. But if Apple could cut back expenses and remain profitable, even with lower sales, it could stay separate from the Osborne-Atari crowd.

The second task was to rescue Lisa. Product success builds on product success. If you have a failure, you lose credibility on your next product introduction. If Lisa remained a problem, retailers and other people would keep worrying about Lisa, diverting attention from Macintosh. To revive Lisa, Sculley and Jobs put together a plan to cut Lisa prices, expand the number of retailers selling Lisa, and develop some form of compatibility between Lisa and Mac, so that the two computers could form a coherent product family.

While this was going on, Apple began to get some good news in the marketplace. The latest version of the Apple II, called the Apple IIe, was selling like gangbusters. During 1983, Apple sold more than 700,000 Apple II computers, up from approximately

300,000 in 1982. Apple marketing managers began to realize the Macintosh might not make the Apple II obsolete after all. Year after year, sales of the Apple II continued to rise. IBM was already in the market and Mac was in the wings, but people kept buying the Apple II. Apple marketeers began to believe there would be life after life for the Apple II.

Macintosh and the Apple II, once seen as competitors within Apple, began to differentiate themselves. A few years earlier, Apple managers figured the Macintosh and the Apple II would both sell at price points between $1,000 and $1,500. But the price points were now drifting further apart. While the price of the Apple II continued falling, the development costs of Macintosh continued to rise. Macintosh would have to sell at $2,000 or $2,500, well above the Apple II price range.

The target markets for the Apple II and the Mac also began to split. More and more, the Apple II was going toward vertical markets. Since the introduction of the Apple II in 1976, software developers had written thousands of specialized programs for the computer, so it could be used in thousands of specialty applications. There were Apple II programs that could help manage pig farms and others to help educate kindergarten children. Even after the introduction of Mac, these applications would not disappear. Macintosh, meanwhile, would be aimed at the horizontal market of knowledge workers.

About the same time, the Macintosh project received some help from a most unexpected source: IBM. After two years of flawless performance in the personal-computer market, IBM began to make a few mistakes.

For one thing, the IBM Peanut, officially called the PCjr, fell short of expectations. With its toy-like keyboard and limited memory, the PCjr disappointed retailers and customers. People began to realize IBM was not invincible. It was not an automatic winner. Among retailers and software designers, there was a new hesitancy, an uncertainty, about IBM. The computer giant would have to reestablish its credibility in the personal-computer market.

What's more, the PCjr flop gave Apple more flexibility in the positioning of Macintosh. Had PCjr been a raging success, people inevitably would have compared Mac to PCjr, even though they were totally different machines aimed at very different markets. Positioning Macintosh properly would have been difficult. When PCjr fizzled, that problem disappeared.

Even before the PCjr flop, some retailers and software developers had begun to worry about their relationship with IBM. Many retailers had become heavily dependent on IBM. At some stores, IBM products accounted for 75 percent of all sales. No retailer likes to be that dependent on a single supplier.

The retailers' worst fears were realized when IBM began expanding its own distribution channels. It added more IBM Product Centers and expanded its direct sales force. Analysts reported IBM was expected to sell 60 percent of its computers through its own sales channels, up from 40 percent in 1982. Independent retailers were justifiably nervous. What was to stop IBM from selling 70 percent or 80 percent or even 100 percent of its products through its own channels, leaving little or nothing for independent retailers?

Independent software developers faced a similar dilemma. Like retailers, they were becoming heavily dependent on IBM. They made most of their money selling programs to run on IBM personal computers. These software designers had helped make the IBM PC a success, but now IBM was beginning to compete with them. IBM was publishing more and more software itself, taking sales away from software companies. Some software companies licensed their software to IBM, but that was a mixed blessing. They got the licensing fees, but lost the market.

To make matters worse, IBM was reportedly working on a new operating system. By keeping parts of the operating system secret, IBM could make life difficult for independent developers of application programs. Major software companies were clearly feeling uneasy about the trends in the industry. Fred Gibbons, president of Software Publishing, the developer of the PFS family of software products, told me he was "sweating bullets" over how to deal with the changing business conditions.

All of this activity added up to a new environment in the personal-computer industry. The concept of Fear, Uncertainty, and Doubt had been turned on its head. In most markets, IBM uses FUD to its advantage. Customers are fearful of buying from any supplier other than IBM. It is the safe bet in an uncertain world.

But in the personal-computer market, IBM had FUDed itself. The company itself was the source of uncertainty and doubt. Members of the industry infrastructure—retailers, software designers—no longer trusted IBM. They were suspicious of IBM's motives, uncertain of IBM's future directions. They still wanted to do business with IBM, but they didn't want to be too dependent on the computer giant.

Macinfrastructure

This new environment presented a tremendous opportunity for Apple and Macintosh. Members of the industry infrastructure no longer saw the world through Big Blue-tinted glasses. They were looking for alternatives to the IBM PC. Apple recognized this, and set out to turn members of the infrastructure into Mac believers.

The industry infrastructure is enormously important in the personal-computer business. No personal computer, no matter how powerful it is, no matter how advanced it is, can win in the marketplace without the support of the infrastructure. Software designers must write programs for the computer, retailers must carry it on their shelves, analysts must praise it in their newsletters.

If a product can win the support of the infrastructure, it is almost certain to win in the marketplace. The infrastructure works like a chain reaction. News about the product spreads by word of mouth, and enthusiasm grows. If programmers write software for the new machine, retailers are more likely to carry it. If more retailers carry it, analysts and journalists tout it as a winner. The product builds momentum and credibility. For customers confused by new technologies and changing markets, the computer looks like a safe bet.

Months before the Macintosh introduction, Apple began working on the infrastructure. As word spread about IBM's plans to sell 60 percent of its computers through its in-house channels, Apple solidified its own relations with dealers. It set up regional dealer councils to serve as a liaison between itself and the retailers. And it cut back its plans for a direct-sales force, making a commitment to independent dealers that they would remain the primary sales channel for Apple computers.

Apple also went after software developers. Here it had some fences to mend. At one time, Apple had been the darling of the software industry. Everyone had wanted to design software for the Apple II. But Apple had turned arrogant with success. When it developed the Apple III and Lisa, Apple didn't let software companies work with the computers before the product introduction. Software designers grew frustrated with Apple, especially when the Apple III and Lisa fell short of expectations. Software companies began focusing on the IBM PC instead.

Now, Apple tried to turn that around. With new humility, Apple managers and engineers visited software developers and asked them to develop programs for the Macintosh. It offered to help and support them in their development efforts. About 100 companies signed up, including three of the biggest and most influential—Microsoft, Lotus, and Software Publishing.

Each software company signed a nondisclosure agreement. That didn't stop the word from spreading. The software community is small, and everybody talks to everybody else. Before long, everybody was talking about the Mac. Designers heard that Microsoft and Lotus were working on Mac software, so they assumed Mac must be a winner. They wanted to get in on the action too. Everybody wanted to design software for the Mac.

Dealers like to hear that type of commitment directly from top management, so John Sculley travelled around the country and met with all the Apple dealers. By the time of the introduction, about 4,000 dealers had been trained on Macintosh. Many had fallen in love with it. Apple won shelf space for the Mac. More important, it won space in the dealers' minds.

Finally, Apple took its message to analysts and industry luminaries, and, later, to journalists. Key members of these communities got seven-hour demonstrations of the Macintosh, with plenty of hands-on time. Many of these people are computer afficianados, and they fell in love with the Mac as soon as they began playing with it. Before long, they started to spread the good word about the new product.

The mission was complete: The infrastructure was lined up solidly behind Macintosh.

Macmessage

In preparing its merchandising and public-relations campaigns, Apple marketeers had one overriding goal: They wanted to establish Macintosh as the third standard in the personal computer industry.

Apple argued that only two products had emerged as industry standards in the eight-year history of the personal-computer industry. Those products were the Apple II and the IBM PC. The Mac would not become a new standard overnight, but Apple wanted to plant the idea early.

To convince customers and the media that Mac was indeed a new standard, Apple stressed its product features. Mac marketeers wanted to drive home the point that Mac was radically different from other personal computers. They identified four key messages about the Mac. Then they repeated those messages over and over. The messages were:

Mac offers "Lisa technology." Although Lisa had fallen short of expectations in the marketplace, its technology won rave reviews. People were intrigued and impressed with Lisa's "friendly" interface—its mouse pointer, its pull-down menus, its bit-mapped graphics, its windowing capabilities. Apple wanted people to know they could get the same features in Macintosh. Or, as Apple marketeers love to say, Mac offers "radical ease of use."

Mac uses a 32-bit processor. Many people don't understand what "32-bit" means, but they know that it stands for advanced technology. After all, the processor in the IBM PC is only a 16-bit processor. As one piece of Mac literature put it: Mac offers "incredible power under the hood."

Mac offers personal-productivity tools. This is where Apple's infrastructure development paid off. At the time of the Mac introduction, Apple could boast that 100 leading software firms were working on Mac software. That software would increase personal productivity and creativity.

Mac comes in one box. The message helped eliminate customer fear. Mac is compact and simple. You can take it out of the box and plug it in. If you want to move the computer, it is easy to carry around. As the marketing plan stated: "Macintosh fits comfortably on your desk and in your life."

The Mac marketeers used these four messages everywhere: in meetings with the media, in meetings with dealers, in customer brochures. At the time of introduction, every one of the 10,000 salespeople selling the Macintosh could recite the four key messages. Apple kept its messages clean and simple. With the Apple IIe, Apple gave forty pieces of information to dealers. With Mac, the company gave dealers a single book.

In meetings with journalists, Apple added a number of other messages about Macintosh. For some publications, Apple stressed the new automated Macintosh factory. The factory, filled with the latest robotics equipment, would turn out one Macintosh every twenty-seven seconds. The story of the factory fit in well with the market environment. Many people were worried about the manufacturing capabilities of American companies. The United States seemed to be losing out to Japan in robotics and other manufacturing technologies.

The Mac factory stood out as a bright spot in this dark environment. While Atari had just shifted its manufacturing overseas, Apple was bringing its manufacturing back to the United States. What is more, the Mac factory was in Fremont, California, where General Motors had just closed down an automobile factory. It was a clear case of a new industry taking over from an old one. It helped give Macintosh a higher profile. Mac was not just another computer. It was a symbol of the American future.

The Mac marketeers also focused on the engineering team that developed the Mac. The team consisted of a dozen or so young people who had contributed their sweat and talent for Macintosh. They had worked day and night for four years. The story of the engineering team showed Apple as a human company, a personal company.

Apple already had a strong corporate personality based on the story of Steve Jobs. The Macintosh story built on that image. Apple was a company for which the public liked to root. Apple people were young, dynamic, and innovative. IBM, on the other hand, was perceived as an anonymous, monolithic corporation. Millions of people knew the story of Steve Jobs. How many people even know the name of IBM's president?

With these and other stories, Apple turned Macintosh into a huge media event. Mac managers began giving key journalists a sneak preview of the machine months before its January 24 introduction. In mid-January, they went on a big press tour, capped by a Macintosh "coming-out party" on January 22. The day of the introduction, Apple mailed out 3,000 press kits, each containing not only pictures and press releases, but a Macintosh T-shirt. After five years of work, Macintosh finally was moving outside of Apple and into the world.

Macadvertising

Apple's advertising strategy for Macintosh broke into two categories: the "1984" television ad and everything else.

The "1984" ad, created by Chiat/Day Advertising, used images from George Orwell's classic book. It was unlike any ad that Apple—or any other company, for that matter—had ever done. It was shown nationally only once, during the 1984 Super Bowl. But it became the most talked-about advertisement in years. Inside Apple, it caused arguments, controversy, and, in the end, big smiles.

The ad looks like a clip from a movie or a rock video, not a television commercial. It begins with a view of a dark and somewhat eerie room. Men with shaved heads sit on row after row of benches. They stare blankly at a huge screen on the wall, where a cold and grim man, clearly representing Big Brother, talks in a monotone voice.

Suddenly, the camera shifts to a young woman dressed in bright red running shorts and an Apple shirt. She is running down a dark corridor, carrying a sledgehammer. Chasing her is a group of uniformed men, apparently the Thought Police. As she enters the main room, she swings the sledgehammer and throws it at the screen. The screen shatters and a huge gust of wind sweeps past the zombie-like men.

The screen goes blank, then the Apple logo appears. A narrator says: "On January 24, Apple will introduce Macintosh. And you'll see why 1984 won't be like *1984.*"

The ad, produced in London for $500,000, almost died before it reached the air. Mike Murray, Mac's marketing manager, showed the commercial to Apple's board of directors in November. Murray loved the ad and expected the board would love it too. He was in for a surprise. Murray explains it this way:

---

*W*hen the commercial was over, I thought I had just come out of a funeral. Their faces were extremely solemn. I thought that I had just made a major career decision. One of the directors looked at Steve (Jobs) and said: "You really like that?" He was just incredulous. Another director had his head on the table and was pounding with his fist. At first, I thought he was laughing. But he was nearly crying.

Murray tried to explain the reasoning behind the ad. He believed that Apple needed an attention-grabber. People in the personal-computer industry already knew about Macintosh. But out in the marketplace, Macintosh was a total unknown. If nothing else, the ad would establish the Macintosh name. Even if the commercial seemed a bit bizarre, or maybe because it seemed a bit bizarre, many of the 80 million people watching the Super Bowl would remember the Macintosh name. The commercial would make them sit up and take notice: Macintosh was something new and radically different.

In addition, the commercial would add to Apple's personality. It showed Apple as daring and creative, and it set the stage for the Apple-IBM battle. Many viewers would recognize Big Brother as a thinly-veiled image of IBM. The interpretation was clear. Apple, the daring and creative upstart, was taking on the colossus. Framing the battle that way gave Apple a big advantage. Americans like to root for underdogs. Everyone likes to see entrepreneurs succeed. It made sense for Apple to play up the Apple-IBM battle.

The board didn't buy these arguments. They told the Mac managers to resell the air time they had bought. Apple tried to sell the air time, which cost $950,000 per minute. But as luck would have it, Apple couldn't find any buyers, and the "1984" commercial went on the air.

Reaction to the commercial was extraordinary. Everyone in the advertising community was talking about it. Newspapers ran articles about it. Television news programs showed the commercial as a news item, giving the Macintosh even more exposure. Inside Apple, the doubters and skeptics, myself included, realized the commercial had worked. At the next board meeting, Murray got a standing ovation.

The rest of the Mac advertising took a very different approach. If the "1984" commercial appealed at an emotional level, the rest of the advertising appealed at a rational level. The ads, both print and television, were very product oriented. They presented product features and product benefits. And they made direct comparisons between the Mac and the IBM PC.

This was a switch from Apple's advertising for Lisa. Apple had used lifestyle ads to promote Lisa. One television ad showed a young executive, dressed in running clothes, working with his Lisa in his office. He was on the telephone with his wife, telling her he would soon be home for breakfast.

Apple learned a lesson from those ads. It learned that lifestyle ads simply do not work for complex new products. With new technologies, differentiation must begin with the product. Companies must start by giving tangible evidence of the product benefits. Companies can't just go out and say: We are the leaders. Intangibles, such as leadership image, must grow out of the tangibles. Corporate and lifestyle advertising start at the wrong end. They start with the intangibles.

With its Macintosh advertisements, Apple shifted back to tangibles. A big chunk of the advertising budget went toward a twenty-page magazine insert, which ran in ten different magazines including *Fortune*, *Time*, and *Business Week*. The insert was filled with facts and figures about Macintosh. On one page, a cut-away diagram showed the inside of a Macintosh. On a four-page foldout, the ad showed how to use the Macintosh mouse. Point. Click. Cut. Paste.

The insert was enormously successful. It had one of the highest recall rates of any advertisement ever run. Apple printed up extra copies of the advertisement and used it as a point-of-sale brochure. The ad did its job well: It helped differentiate Macintosh from other computers and it solidified Macintosh's product positioning.

Macfuture

Apple hoped to sell 50,000 Macintoshes during the first 100 days following the product introduction. In fact, it sold more than 75,000. Clearly, the marketing plan was working. Macintosh had achieved a unique presence in the market.

As is often the case, Macintosh's product position cannot be articulated in a short sentence. There is no single slogan. Rather, Macintosh's positioning arose from a variety of factors: Customers see the product as technologically advanced and easy to use; dealers want to carry it; software people enjoy working with it; it is produced in the only automated personal-computer factory in the world; it was designed by a bunch of creative young engineers. All these things add up to a product position.

The marketing work on Macintosh is hardly done, however. The product introduction was just the beginning. In many ways, the product introduction is the easiest thing to do in marketing. It's like having a baby. Giving birth seems tough at the time, but raising the child is a lot tougher. Growing is a process of change, experimentation, and adaptation. Parents have to provide guidance, support, and discipline. The same goes for products.

Apple still has a lot to learn about marketing. Apple is a great product development company. There is nobody better in the computer industry. And Apple is a great promotions company. It can throw a great party and get a lot of great press. But in between product development and promotion, there is a lot of marketing to be done. For example, marketing managers must decide on target markets and figure out how to offer solutions to these markets.

Marketeers at Apple still don't have a very good handle on Macintosh's target audience. Macintosh is seen as a trendy, upscale product. It is being bought by a subset of knowledge workers—the knowledge yuppies. Mac marketeers must figure out ways to reach the rest of the 25 million knowledge workers. The idea of an "information appliance" hasn't really caught on. Apple must find new ways to communicate its message about Macintosh.

The marketing of Macintosh evolved quite a bit in the three years before the introduction. It will continue to evolve over the next few years. In many ways, the early Macintoshes are development tools. Apple is pushing Macs onto the market, and it is waiting for the market response. Customers are giving feedback to Apple. Adaptation will follow.

In reaction, Apple will experiment with new strategies and it will modify the product. Constant adjustment is the only way to maintain a position in a dynamic marketplace. Less than a year after the Macintosh introduction, Apple introduced a laser printer and networking capabilities to make Macintosh better suited to office applications. There will be more changes in the future. More memory will be added. New peripherals will be added. Color will be added. New video capabilities will be added. Eventually, specialized software to reach specific types of knowledge workers will be offered.

While making all these changes, Macintosh managers must avoid the pitfalls of bigness. In the development of Mac, intuition reigned supreme. There was little market research and little managing. Everyone was close to the product. As the Macintosh division grows, managers will inevitably become more detached. They will spend more time managing. They won't have time to walk around and talk to people in the lab.

Can Macintosh continue its success? Only time will tell. Nothing is certain or fixed in the computer industry. The marketplace will continue to change, technologies will continue to change— and Macintosh will continue to change. That is the way of life in dynamic industries.

Index

1-2-3, 67
 and qualitative approach, 28
3Com, and press relations, 84-85
90/10 rule, 60
 and press relations, 86

Acquisition strategies
 and AM International, 71
 and Exxon Corporation, 70-71
 and Schlumberger, 70
 and Western Electric, 71
Activision, and market environment,
 111
Adaptation sequence, 79-82
 and General Electric, 81
AM International, and acquisition
 strategies, 71
Antitrust, and strategic
 relationships, 76
Apple Computer
 and dynamic positioning, 23, 25
 and market environment, 108
Apple II
 history of, 153
 and Macintosh, 156, 159, 163
 and market environment, 31-33
 and market perception, 140
Apple IIe, and Macintosh, 162-163
Apple III, 53
ASK Computer, and creative
 segmentation, 79
Avis-Hertz rivalry, and dynamic
 positioning, 14

Bell, Gordon, 67
Bigness mentality, 117

and IBM, 142
and intangible competition,
 140-142
Boone, Gary, 132-133
Burroughs, and Convergent
 Technologies, 55

Cohesive Network Corp.
 and external audits, 116
 and internal audits, 102
Commodity mentality, and
 intangible competition,
 140-142
Compaq, and Ben Rosen, 55
Convergent Technologies, and
 Burroughs, 55
Convex Systems
 and external audits, 116
 and positioning strategy, 120-122
 and strategic customers, 78
Corporate positioning
 benefits of strong, 93-94
 factors in, 91-92
 and Inmos, 96
 and Intel, 95-96
 and the Japanese, 92-93
 loss of, 95
 and silver bullets, 93
 and Tandem Computer, 94-95
 and Trilogy Systems, 95
Couch, John, 159
Creative segmentation, 78-79
 and ASK Computer, 79
Customer, and intangible
 competition, 79

Digital Research, Inc.
  and strategic relationships, 74-75
  and targeting markets, 46-47
Drori, Zeev, 37
Drucker, Peter, 59, 109
Dynamic positioning, 14-19
  and Apple, 23, 25
  and Avis-Hertz rivalry, 14
  and the infrastructure, 16
  and Intel, 24
  key ideas of, 19-33
  and market environment, 15
  and market structure, 15
  stages of, 15
  and strategic relationships, 15
  and Tandy, 23

Eli Lilly, and Genentech, 55
External audits
  and Cohesive Network, 116
  and Convex, 116
  and Intel, 116
  and market environment, 108
  procedure for, 114-115
  and Rolm, 115-116
Exxon Corporation, and acquisition
    strategies, 70-71

FUD, 20, 164-165
  and IBM, 54
Focus groups, and qualitative
    approach, 113

Gelbach, Ed, 106
Genentech
  and Eli Lilly, 55
  and strategic relationships, 75-76
General Electric, and adaptation
    sequence, 81
Genetic Systems, and Syntex, 56
Gibbons, Fred, 164
Gravitational forces, 19-21
Grove, Andy, 40, 44, 124

Hewlett, Bill, 103
Husserl, 119

IBM
  and the bigness mentality, 142
  and FUD, 54
  and intangible positioning factors,
    44

and Intel, 70
and Macintosh, 70, 160-161,
    163-165
and Microsoft, 69-70
and the PCjr, 25
and Sytek, 55
and Tandon, 70
Imagic, and positioning strategy,
    120
Infrastructure, 21005
  and dynamic positioning, 16
  and Intel, 67
  and Macintosh, 165-167
  and market positioning, 61-67
  and National Semiconductor, 65
  and press relations, 85
Inmos, and corporate positioning,
    96
Intangible competition
  and the bigness mentality,
    140-142
  and broken chains, 142-145
  and change, 130-132
  and the commodity mentality,
    138-140
  and the customer's mind,
    136-138
  definition of, 129-130
  and product concept, 145-146
  and psychological bogeymen, 137
  and public knowledge, 134-136
  and resistance to change,
    132-134
  and surprises, 146-148
  and you, 148
Intangible positioning factors
  description of, 16
  and IBM, 44
  and Intel, 43
  and product positioning, 41-44
  and Zeiss, 43
Intel
  and corporate positioning, 95-96
  and dynamic positioning, 24
  and external audits, 116
  and IBM, 70
  and the infrastructure, 67
  and intangible positioning factors,
    43
  and positioning strategy, 122-125
  and product positioning, 40

and resistance to change, 134
Internal audits
  and Cohesive Network Corp., 102
  conflict in, 105-106
  cultural factors in, 102-103
  and National Semiconductor,
    103-104
  and qualitative information,
    104-105

Jaunich, Robert, 26
Jobs, Steve, 4, 31, 133, 147, 154,
  159, 160, 169

Kapor, Mitch, 28

Levitt, Theodore, 40, 102, 139
Levy, James, 111
Lisa
  and Macintosh, 158, 162
  reception of, 159-160
Litronix, and press relations, 84
Lotus
  and 1-2-3, 67
  and Ben Rosen, 55, 67

Microelectronics and Computer
  Technology Corporation, and
  strategic relationships, 76
Macintosh
  advertising strategy for, 169-172
  and Apple II, 156, 159, 163
  and Apple IIe, 162-163
  audience for, 154-156
  and the future, 172-173
  and the future of Apple, 162
  history of, 152-154
  and IBM, 160-161, 163-165,
    170-171
  and the infrastructure, 165-167
  and Lisa, 158, 162
  marketing of, 151-162
  and Peanut, 161, 163
  and public knowledge of, 135
  and public relations, 167-169
  and turmoil at Apple, 157-162
Market-creating strategies, 21-25
Market-driven approach, 26
Market environment
  and Activision, 111
  and Apple Computer, 108
  and the Apple II, 31-33

definition of, 29
  and dynamic positioning, 15
  and external audits, 108
  and product positioning, 39
Marketing-driven approach, 26
  and Rolm, 26
  and TeleVideo, 26
Market positioning
  and the infrastructure, 61-67
  and the press, 82-88
  and strategic relationships, 68-77
  strategy for, 57
  and word of mouth, 57-61
Market-share mentality, 21-25
Market structure, and dynamic
  positioning, 15
Measurex, and product positioning,
  40-41
Metaphor Computer Systems, and
  targeting markets, 45-46
Microsoft, and IBM, 69-70
MiniScribe, and strategic
  relationships, 77
Monolithic Memories, and product
  positioning, 37
Moore, Gordon, 44, 124
Morgan, James, 26
Morgan, Jim, 27, 56
Murray, Mike, 157, 170

National Semiconductor
  and the infrastructure, 65
  and internal audits, 103-104
Noyce, Bob, 44, 124

Packard, Dave, 103
PCjr, 25. *See also* Peanut
Peanut, and Macintosh, 161,
  163-164
Positioning sessions
  goals of, 118
  procedures for, 118-120
Positioning strategy
  and Convex, 120-122
  developing, 101
  and Imagic, 120
  and Intel, 122-125
  re-evaluating, 122
Press relations
  and 3Com, 84-85
  and the 90/10 rule, 86

and the infrastructure, 85
and Litronix, 84
and market positioning, 83-88
and Synapse Computer, 84
Procter & Gamble, and public
  knowledge of, 135
Product-development cycle, stages
  of, 71-73
Product positioning
  definition of, 37
  and intangible positioning factors,
    41-44
  and Intel, 40
  key ideas of, 37-38
  and market environment, 39
  and Measurex, 40-41
  and Monolithic Memories, 37
  and Vitalink, 49
Psychological bogeymen, and
  intangible competition, 137

Qualitative approach
  and 1-2-3, 28
  definition of, 28
  and focus groups, 113
  and Spectra-Physics, 114
  and Tektronix, 110
Qualitative information, and
  internal audits, 104-105

Raskin, Jeff, 152-154
Regis McKenna, Inc., 3, 8
Reich, Robert, 142
Resistance to change, and Intel, 134
Rolm
  and external audits, 115-116
  and marketing-driven approach,
    26
Rosen, Ben, 61-62, 65
  and Compaq, 55
  and Lotus, 55, 67

Schlumberger, and acquisition
  strategies, 70
Sculley, John, 26, 113, 162, 166
Silicon Graphics, and targeting
  markets, 45
Silver bullets, and corporate
  positioning, 93
Spectra-Physics, and qualitative
  approach, 114
Sporck, Charlie, 103

Strategic customers
  and Convex Systems, 78
  and strategic relationships, 77-78
  and Valid Logic, 78
Strategic relationships
  and antitrust, 76
  and Digital Research, Inc., 74-75
  and dynamic positioning, 15
  and Genentech, 75-76
  and Japan, 73
  and market positioning, 68-77
  and Microelectronics and
    Computer Technology
    Corporation, 76
  and MiniScribe, 77
  and strategic customers, 77-78
  and ZyMos, 75
Synapse Computer, and press
  relations, 84
Syntex, and Genetic Systems, 56
Sytek, and IBM, 55

Tandem Computer, and corporate
  positioning, 94-95
Tandon, and IBM, 70
Tandy, and dynamic positioning, 23
Targeting markets
  and Digital Research, Inc., 46-47
  and Metaphor Computer Systems,
    45-46
  reasons for, 45
  and Silicon Graphics, 45
  and TeleVideo, 47
Tektronix, and qualitative approach,
  110
TeleVideo
  and marketing-driven approach,
    26
  and targeting markets, 47
Toffler, Alvin, 5, 136-137
Trilogy Systems, and corporate
  positioning, 95

Valid Logic, and strategic
  customers, 78
Vitalink, and product positioning,
  49

Webster, Jr., Frederick, 7
Western Electric, and acquisition
  strategies, 71

Word of mouth, and market
  positioning, 57-61
Wozniak, Steve, 31, 133, 160

Zeiss, and intangible positioning
  factors, 43
ZyMos, and strategic relationships,
  75